PRAISE FOR *USING SPSS*

This primer is an excellent resource for students learning the basic functions of SPSS, the power of statistics, and the basic premises underlying inferential and descriptive statistics.

—Robert J. Eger III, *Florida State University*

I love this book. . . . The structure, content and format are superior in every way to what I have been using, and will make my task easier and the students learning curve shorter.

—G. L. Forward, *Point Loma Nazarene University*

This book provides a tremendous amount of detail and is useful in walking a new user through the steps of using the software. The visuals are helpful and clear, the level of writing is good for beginning students, and the concepts are presented clearly.

—Poco D. Kernsmith, *Wayne State University*

The key strength of the text is in the opening chapters, which should help students get a strong and confident beginning in the use of SPSS. . . . The text will also be a valuable reference guide for our students as they conduct research in the two methods courses that follow statistics.

—David S. Malcolm, *Fordham University*

I frequently have graduate students that need [help with] SPSS. This text would bring joy into their lives.

—Darlene A. Thurston, *Jackson State University*

USING
SPSS

I dedicate this book to my son, Randy Cunningham, and my friend, Glenn Bailey.

—James B. Cunningham

I dedicate this textbook to my three children, Sally, James (1965–1996), and Wendy.
The encouragement and support for their father and his educational
pursuits was (and is) above the call of duty.

—James O. Aldrich

An Interactive Hands-On Approach

James B. Cunningham ■ James O. Aldrich

California State University, Northridge

Los Angeles | London | New Delhi
Singapore | Washington DC

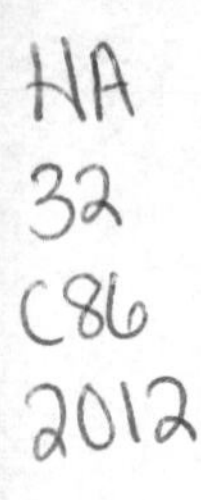

SAGE

Los Angeles | London | New Delhi
Singapore | Washington DC

FOR INFORMATION:

SAGE Publications, Inc.
2455 Teller Road
Thousand Oaks, California 91320
E-mail: order@sagepub.com

SAGE Publications Ltd.
1 Oliver's Yard
55 City Road
London EC1Y 1SP
United Kingdom

SAGE Publications India Pvt. Ltd.
B 1/I 1 Mohan Cooperative Industrial Area
Mathura Road, New Delhi 110 044
India

SAGE Publications Asia-Pacific Pte. Ltd.
33 Pekin Street #02-01
Far East Square
Singapore 048763

Acquisitions Editor: Vicki Knight
Associate Editor: Lauren Habib
Editorial Assistant: Kalie Koscielak
Production Editor: Brittany Bauhaus
Copy Editor: QuADS Prepress (P) Ltd.
Typesetter: C&M Digitals (P) Ltd.
Proofreader: Scott Oney
Indexer: Rick Hurd
Cover Designer: Bryan Fishman
Marketing Manager: Helen Salmon
Permissions Editor: Adele Hutchinson

Printed in the United States of America

Library of Congress Cataloging-in-Publication Data

Cunningham, James B., 1938-

Using SPSS: an interactive hands-on approach / James Cunningham, James O. Aldrich.

p. cm.
Includes bibliographical references and index.

ISBN 978-1-4129-9515-3 (pbk.)

1. SPSS (Computer file) 2. Social sciences—Statistical methods—Computer programs. I. Aldrich, James O. II. Title.

HA32.C86 2012
005.5′5—dc23
2011019175

This book is printed on acid-free paper.

11 12 13 14 15 10 9 8 7 6 5 4 3 2 1

Brief Contents

DETAILED CONTENTS

About the Authors

James B. Cunningham (PhD in Science Education, Syracuse University) is Professor Emeritus of Science and Computer Education and former chair of the Department of Secondary Education at California State University, Northridge. Formerly chair of the Departments of Science and Mathematics in Washington State high schools, he is author of *Teaching Metrics Simplified* and coauthor of *BASIC for Teachers*, *Authoring Educational Software*, *Hands-On Physics Activities With Real-Life Applications*, and *Hands-On Chemistry Activities With Real-Life Applications*. He used SPSS extensively during his tenure as director of the Credential Evaluation Unit in the College of Education. He is currently a Fellow in the Center for Teaching and Learning at California State University, Northridge.

James O. Aldrich (Doctor of Public Administration, University of Laverne) is a retired lecturer on statistics and research methods at California State University, Northridge. He has served as the principal investigator and codirector of a National Cancer Institute–funded research project. He held the appointment of instructor in the Department of Pathology at the University of Southern California School of Medicine. He has served on various committees for the Los Angeles chapter of the American Statistical Association and has also taught biostatistics, epidemiology, social statistics, and research methods courses for 20 years. The primary statistical software used for his coursework has been SPSS.

Preface

Using SPSS: An Interactive Hands-On Approach is intended as a self-instructional book in which the readers, be they teachers, students, or anyone, can learn to use IBM® SPSS® (PASW)* 18.0 on their own, and at their own pace, without the assistance of an instructor. As such, it would make a solid companion text for most statistics or research methods courses. The book uses statistics to teach SPSS, not SPSS to teach statistics. Each chapter includes an introduction and list of performance objectives indicating what the reader will be able to do after reading the chapter. SPSS Version 18 for Windows was used in the preparation of this book. SPSS was acquired by IBM in October 2009 and is an IBM Company.

The book is unique in that it encourages the reader to interact with SPSS on the computer as he or she works through examples in each chapter. This approach to learning may be novel to the reader, but the authors feel the best way to learn a subject is to interact with it in a meaningful manner. We have made every effort to ensure the book is "user friendly" as we guide the reader through the interactive learning process. Bulleted phrases provide a step-by-step procedure to be followed by the reader when completing the exercises.

Of course, one can read the book and learn about SPSS without interacting with SPSS on the computer, but the reader will miss the excitement of discovery, and that's where the fun comes in. Learning by inquiry and discovery helps one personalize and remember information and processes, and SPSS is definitely about these.

Using SPSS: An Interactive Hands-On Approach is intended for readers who may have some background in statistics and for those who know little or nothing about statistics but who want SPSS to do the actual statistical and analytical work for them. They want to know how to get their data into SPSS, how to organize and code that data so SPSS can make sense of it, and how to

*IBM® SPSS® Statistics was formerly called PASW® Statistics.

ask SPSS to analyze that data and report out with tables, charts, and graphs. In short, they want SPSS to do the hard work!

The book is structured such that, after Chapter 3, the chapters may be read in any order depending on the needs and interests of the reader. The book is appropriate for most disciplines, including education, sociology, psychology, communications, business, and public administration, because the data sets used are rather generic and are included only for the purpose of describing and explaining the appropriate SPSS statistical procedures associated with these data sets. Some of the data sets are small and are included to minimize the amount of data to be entered by the reader. As such, the number of cases in these data sets does not always meet the requirements of the Central Limit Theorem.

Many of the chapters include screenshots showing the reader exactly how and where to enter data. The material covered in Chapters 1 through 4 provides essential information regarding navigating in SPSS, getting data in and out of SPSS, and determining the appropriate level of measurement required for a statistical test. Chapters 5 and 6 describe additional methods for entering data, validating data, and working with data and variables. Chapter 7 describes and explains the Help Menu available in SPSS and how to find information on various statistical tests and procedures. Chapters 8 and 9 provide hands-on experience creating and editing graphs and charts. Chapter 10 provides explicit directions for printing files, output from statistical analysis, and graphs. Chapter 11 describes and explains basic descriptive statistics. Finally, Chapters 12 through 23 provide hands-on experience employing the various statistical procedures and tests available in SPSS, including both parametric and nonparametric tests. Appendices A, B, C, and D contain data sets the reader will use, information on inferential statistics, and information on the nature of statistical tests, including a description related to stating hypotheses and selecting significance levels.

A novel approach taken in this book is the inclusion of nonparametric statistical tests in the same chapters with their related parametric tests in case the assumptions required for the parametric tests cannot be met. Other books describe parametric and nonparametric tests in separate chapters, which the authors feel is inefficient because it forces the reader to continually move from one section of a book to another in search for the rationale justifying the use of either type of test.

SPSS is an outstanding statistical package. A primary reason for our writing this book is to make the benefits of the SPSS program available, not only to the novice but also to the more experienced user of statistics. The book is appropriate for lower-division, upper-division, and graduate courses in statistics and research methods.

△ HOW TO USE THIS BOOK

As the reader will note in the first lesson in Chapter 1, the authors use a simple format to allow the reader to respond to requests. The reader will be moving the mouse around the computer screen and clicking and dragging items. The reader will also use the mouse to hover over various items to learn what these items do and how to make them respond by clicking on them. Things the reader should click on or select are in **boldface**. Other important terms in the book are in *italics.* Still other items are enclosed in parentheses.

The reader will often be requested to enter information and data while working through the examples and exercises in this book. To help in this procedure, we often open windows on the screen showing exactly, step-by-step, where to enter this information or data from the keyboard. And, at times, we use cutouts (balloons) and screenshots to indicate where to enter information.

ACKNOWLEDGMENTS

I wish to thank my colleagues, Richard Goldman, Wendy Murawski, and Marcia Rea, in the Center for Teaching and Learning at California State University, Northridge, for planting the seed for this book in the authors' mind and for their encouragement while this book was being written. In addition, I wish to thank Michael Spagna and Jerry Nader, Michael D. Eisner College of Education, for their ongoing support.

—*James B. Cunningham*

I first thank my students, who for many years followed my often hastily written instructions on how to get SPSS to do what it was supposed to do. Second, I thank my coauthor, who had the idea for the book and invited me to participate in its writing. I also thank my teaching assistant Hilda Maricela Rodriguez for her careful and tireless review of all the SPSS steps and screenshots presented in the book.

—*James O. Aldrich*

We wish to thank the professionals at SAGE Publications for their invaluable contributions to the publication of this book. If Vicki Knight, Acquistions Editor, had not seen merit in our proposal, this work would not have been possible. Lauren Habib, Associate Editor, Kalie Koscielak, Editorial Assistant, and Brittany Bauhaus, Production Editor, always kept us on track during the editing and production process. Many thanks to Shankaran Srinivasan for excellent copy editing. Bryan Fishman produced some great graphics and a perfect cover for the book. Many thanks to Helen Salmon, Marketing Manager, and Adele Hutchinson, Permissions Editor, for their assistance.

The authors and SAGE would also like to acknowledge the following reviewers for their contributions to this text:

Robert J. Eger III, *The Florida State University*

G. L. Forward, *Point Loma Nazarene University*

Vanessa Jackson, *University of Kentucky*

Poco D. Kernsmith, *Wayne State University*

David S. Malcolm, *Fordham University*

Emily J. Nicklett, *University of Michigan*

Darlene A. Thurston, *Jackson State University*

CHAPTER 1

FIRST ENCOUNTERS

1.1 INTRODUCTION AND OBJECTIVES △

Hi and welcome to SPSS. Assume you know nothing about variables, values, constants, statistics, and those other mundane things. But do assume you know how to use a mouse to move around the computer screen and how to click an item, select an item, or drag (move) an item.

We have adopted an easy mouse-using and typing convention for you to respond to our requests. For example, if you are requested to open an existing file from the SPSS Menu, you will see click **File**, select **Open**, and then click **Data**. In general, we will simply ask you to click an item, to select (position the pointer over) an item, to drag an item, or to enter data from the keyboard. Note that in SPSS, columns in sheets run vertically and rows run horizontally, as in a typical spreadsheet such as Excel.

OBJECTIVES

After completing this chapter, you will be able to

- Enter variables into the Variable View screen
- Enter data into the Data View screen
- Generate a table of statistics
- Generate a graph summarizing your statistics
- Save your data

△ 1.2 ENTERING, ANALYZING, AND GRAPHING DATA

We are going to walk you through your first encounter with SPSS and show you how to enter some data, analyze those data, and generate a graph. Just follow these steps:

- Start SPSS, and a window will open. When you see "What would you like to do?" click **Type in data** and click **OK**. You will see a data screen.
- At the bottom of the screen, click **Variable View**.
- At the top of the screen, enter the word **Height** in the cell (the box below *Name* and to the right of Row 1). The callout (balloon) points to the cell in which you are to enter Height. Cells are those little boxes at the intersection of columns and rows. See Figure 1.1.

Figure 1.1 Variable View screen

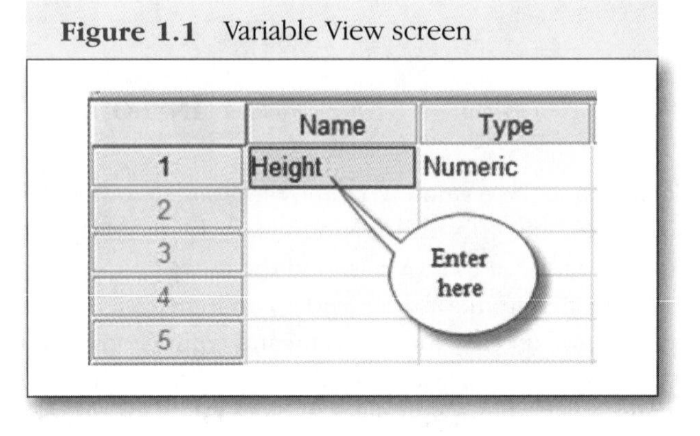

- At the bottom of the screen, click **Data View**.

You will now enter the heights of five persons in inches. Enter the value 40 in Row 1 below Height. The balloon indicates the cell in which you are to enter 40 as shown in Figure 1.2. Click in Row 2 below Height, and

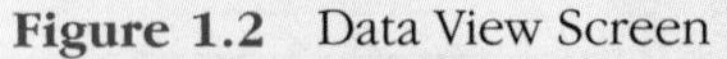

Figure 1.2 Data View Screen

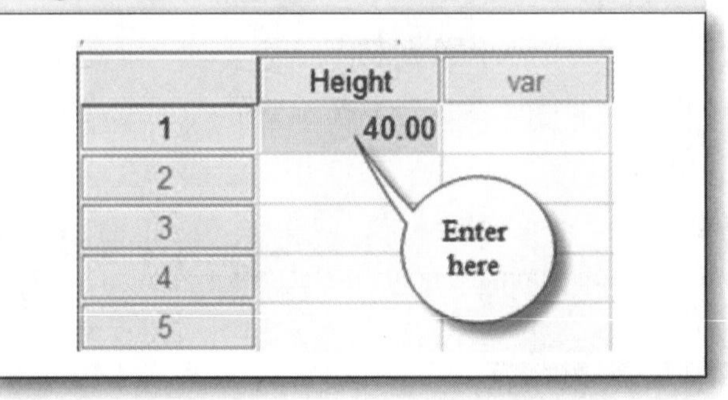

enter 36. Do the same for Rows 3 to 5, entering 38, 41, and 68, respectively. If you make a mistake in entering the numbers, just click the cell and reenter the number.

- After you have entered the five pieces of data, check carefully to see if the entries are correct. If they are, save your work as follows: click **File** and then click **Save As**. A window titled "Save Data As" will open in which you will enter a name for your work (project). You could enter any name you wish, but for this exercise, enter the name **chapter1** in the "File Name" box (block). The "Look in" box showing where the file will be saved should have an entry titled "Documents." Click **Save**. Your data have been saved.

Note: Save your work often when working in SPSS, in case your computer malfunctions or some other problem interrupts your session.

- Let's continue with the exercise. On the SPSS Menu at the top of the screen, click **Analyze**, select **Descriptive Statistics**, and then click **Descriptives**. A window will appear titled Descriptives.
- Drag **Height** to the Variables box or click **Height** if it is not already highlighted and then click the right arrow to place Height in the box. Click **Options**, and another window will appear. Click **Mean** and **Sum** and unclick any other options that may appear, and then click **Continue**. Click **OK**.

Another screen titled "Output SPSS Statistics Viewer" will appear floating over the Data View screen with a table titled Descriptive Statistics showing the results of the analysis. You will see the sum and average (mean) of the heights you entered.

- On the Main Menu, click **Graphs**, select **Legacy Dialogs**, and then click **Bar**. When the small window opens, click **Simple**, and then click **Values of Individual Cases**. Click **Define** and a window opens. Click **Height** and drag it to the "Bars Represent" box or click the right arrow to place Height in that box. Click **OK**. A graph will appear in the same window under the table.
- After you have reviewed the results of the analysis of your data and the graph, you should save the Output-Viewer screen, which contains the table with the results of your analysis and the graph. In the Output-Viewer screen, click **File**, and then click **Save As**. A window will appear. In the "File name" box, enter **chapter1**. Note that the file name is all lower case and does not include any embedded spaces (blanks). The "Look in" box indicates the location where

your file will be saved and should have an entry titled "Documents." Click **Save**. After saving your work, your Output-Viewer screen will remain. Click the little red "x" in the top right corner to make it go away. Don't worry, because you can easily get it back!

You have just used SPSS to analyze some data and provide some statistical results and a graph. You can see, by inspection of the table, that the sum of the heights is 223.0 inches and the average is 44.6 inches. The cases (persons) 1 through 5 are plotted on the *x*-axis (abscissa) of the graph, and the height values are plotted on the *y*-axis (ordinate). Does the graph indicate that all the persons may be children?

Admittedly, the statistical analysis and graph are not that exciting. But it does show you that SPSS is not difficult to use. Of course, you could have used a handheld calculator to do the same analysis in less than a minute. But suppose you had hundreds of variables such as Height and the same number of cases for each! Using a calculator to analyze these data would be a monumental task. But SPSS can do it easily.

If you wish to exit (quit using SPSS) at this time, click **File** and then click **Exit**.

△ 1.3 SUMMARY

In this chapter, you learned how to enter variable names and data. You also learned how to generate a table of statistics and a graph summarizing those statistics. In the next chapter, you will learn to navigate in SPSS. You will be introduced to the Main Menu, the Toolbar editor, and the options available for these. Finally, you will be introduced to the various boxes and windows in SPSS that allow you to enter information regarding your variables.

CHAPTER 2

NAVIGATING IN SPSS

2.1 INTRODUCTION AND OBJECTIVES

As with any new software program you may use, it is important that you are able to move around the screen with the mouse and that you understand the meaning and purpose of the various items that appear on the screen. Consequently, we present a short tour of the Variable View screen, the Data View screen, the Main menu, and the Data Editor Toolbar. You will use these often as you complete the chapters in this book.

OBJECTIVES

After completing this chapter, you will be able to

Describe the Variable View screen and its purpose

Describe the Data View screen and its purpose

Select items from the Main menu and Data Editor Toolbar

Explore the Variable View screen as it relates to entering the following information and commands into boxes and windows regarding variables: *name, type, width, decimals, labels, values, missing, columns, align, measure*

△ 2.2 SPSS VARIABLE VIEW SCREEN

Start SPSS and click **Variable View** at the bottom of the screen. Figure 2.1 shows a portion of the Variable View screen. We have entered the variable Height in the first cell.

Figure 2.1 Variable View Screen

	Name	Type	Width	Decimals	Label	Values	Missing	Columns	Align	Measure	Role
1	Height	Numeric	8	2		None	None	8	Right	Scale	Input
2											
3											
4											
5											

As you will recall from Chapter 1, you were briefly introduced to the Variable View screen when you entered the variable Height. The rows represent variables and the columns represent attributes (properties) and other information that you can enter for each variable. You must provide a name for each variable or SPSS will assign a default, such as var3. It is in Variable View that you enter all your variables and their properties.

You will often be requested to enter information into a cell. Any cell you click is the active cell, displayed in color, indicating it is ready to receive input from the keyboard. In Figure 2.2, you see an example showing a balloon pointing to the cell in which you are to enter the variable Height.

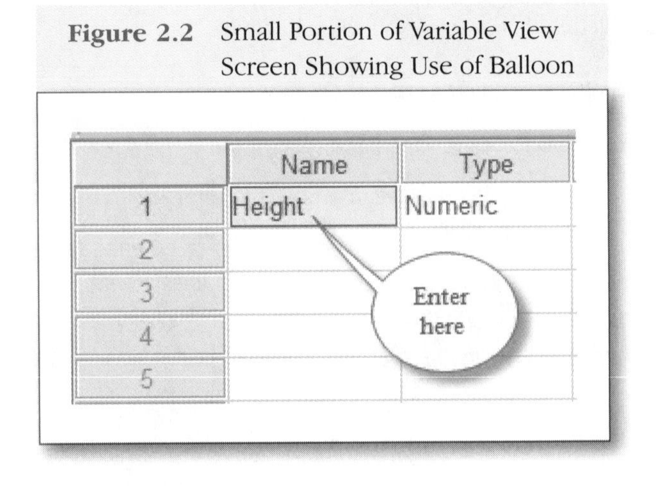

Figure 2.2 Small Portion of Variable View Screen Showing Use of Balloon

2.3 SPSS DATA VIEW SCREEN

A small portion of the Data View screen is shown in Figure 2.3.

Figure 2.3 Data View Screen With Data

	Height	var
1	40.00	
2	36.00	
3	38.00	
4	41.00	
5	68.00	
6		

Click **Data View** if you are not already in that screen. It is in the Data View screen that you enter data for each variable. We have entered the same data in Rows 1 to 5 that you entered in Chapter 1 as shown in Figure 2.3. The Data View screen is similar to the Variable View screen in that it shows rows and columns. However, in Data View, columns represent variables and rows represent cases, also called records, associated with each variable. A record may refer to a student, a teacher, a housewife, or a participant in a pageant, and so on. Although we show only the variable Height for a person in this example, there may be other variables such as weight and gender associated with that person. These variables would be entered and defined in the Variable View screen.

2.4 SPSS MAIN MENU

Figure 2.4 Main Menu

Let's take a quick look at the SPSS Main Menu, referred to hereafter as the Menu, as shown in Figure 2.4. This Menu is displayed at the very top of the Variable View and Data View screens.

- Click **File**, and you will see a drop-down list of options you can choose.
- Click **Edit**, and you will see a list of options available for that item.

- Click the other items on the Menu, one at a time, to see what options are available. If some options are dimmed, this simply indicates that you cannot select these at the moment. You will be introduced to the various items on the Menu and options as you proceed through this book.

△ 2.5 DATA EDITOR TOOLBAR

And now a quick look at the Data Editor Toolbar, shown in Figure 2.5, which is a series of icons displayed horizontally across the page directly below the Menu. If you do not see this toolbar, do the following: On the Menu, click **View**, select **Toolbars**, and then click **Data Editor**.

Figure 2.5 Data Editor Toolbar

If there were no data in the Data screen, some of these icons would be gray, but as soon as any data are entered, all icons will show in color. Put the mouse pointer on the first icon on this toolbar and hover over it. You should see "Open data document," which is asking if you wish to open a document. Place the mouse pointer on the other icons and hover over each so that you can see the purpose of these. Much of what you can do using the Menu can also be done using the Data Editor Toolbar. The toolbar simply makes your work easier by providing a simpler method. Older versions of SPSS may not include all these icons. But those we need are present in every version of SPSS. The Data Editor Toolbar is displayed in both the Variable View and Data View screens unless you choose to hide this toolbar.

△ 2.6 SHORT TOUR OF VARIABLE VIEW SCREEN

We again show a portion of the Variable View screen in Figure 2.6.

Figure 2.6 Variable View Screen

	Name	Type	Width	Decimals	Label	Values	Missing	Columns	Align	Measure	Role
1	Height	Numeric	8	2		None	None	8	Right	Scale	Input
2											
3											
4											
5											

Let's take a quick tour to see what options are available in SPSS for describing and defining variables such as the variable Height. Think of a variable as a container that can hold values. To see how you can enter information regarding variables,

- Click **Variable View**.
- Click in Row **1** below *Name*, and enter the variable Height.
- Click the cell below *Type*. If you click in the left part of the cell you will see a colored square.
- Click the square, and a window will open as shown in Figure 2.7. (Note: If you click in the right part of the cell, the window will open directly.)

Figure 2.7 Variable Type Window

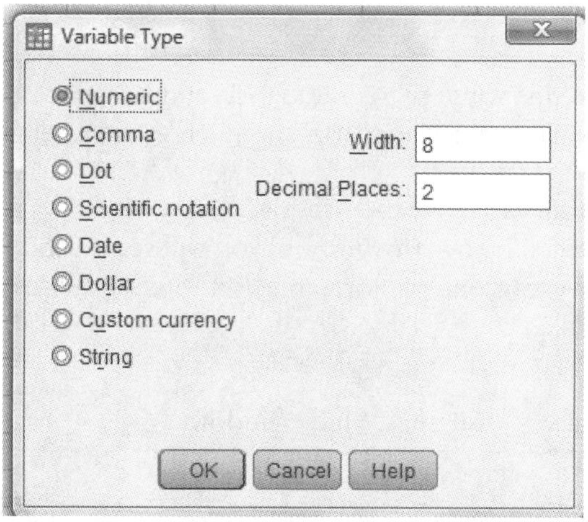

You can select certain settings to tell SPSS what type of numbers or information you wish to enter.

- Click **Cancel** to close the window.
- Click the cell below *Width*. You can use the up-down arrows to set the width of a cell.
- Click the cell below *Decimals*. You can use the up-down arrows to change the number of decimal points in numbers.
- Click the cell below *Label*. You can enter a longer identifying name for a variable.
- Click the cell below *Values*, and a window will open as shown in Figure 2.8.

Figure 2.8 Value Labels Window

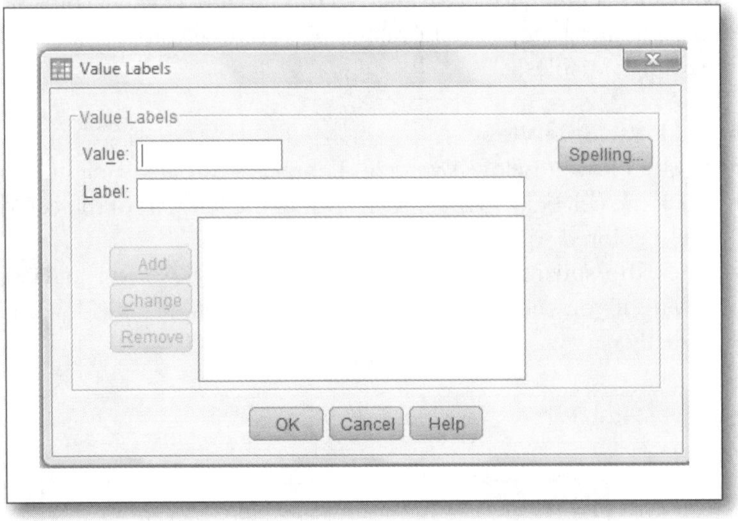

You can use this window to enter labels and values for variables. We will describe and explain values and labels in much greater detail later.

- Click **Cancel** to close the window.
- Click the cell below *Missing*, and you will see a window, Figure 2.9, in which you can enter information on missing values associated with the variables.

Figure 2.9 Missing Values Window

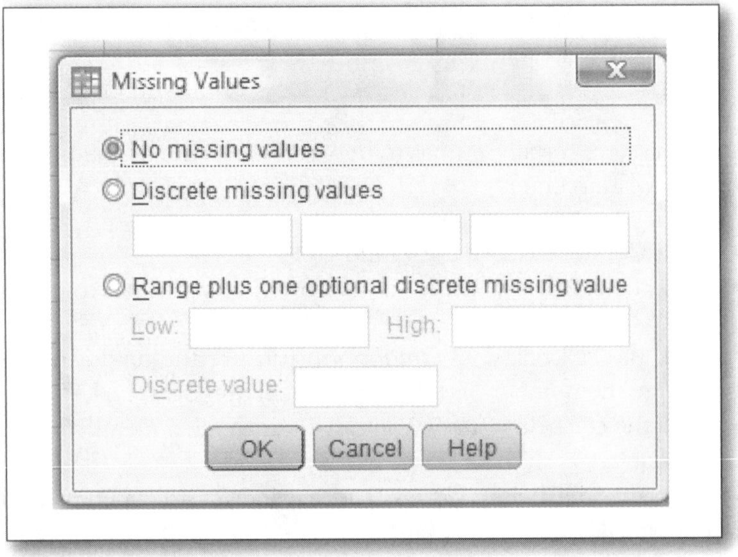

You can use this window to tell SPSS how to handle missing values when doing an analysis associated with variables.

- Click **Cancel** to close the window.
- Click the cell under *Columns*. You can use the up-down arrows to set the width of a column.
- Click the cell below *Align*. You can use the arrow to align information in a column.
- Click the cell below *Scale*. You can use the arrow to indicate whether the level of measurement for a variable is Scale, Ordinal, or Nominal. We will have much to say about these values later.

2.7 SUMMARY △

In this chapter, you have learned to navigate the Variable View and Data View screens and learned that all variables must be entered in the Variable View screen, and all data must be entered in the Data View screen. You were introduced to boxes and windows for entering information regarding variables, including *name*, *type*, *width*, *decimals*, *label*, *values*, *missing*, *columns*, *align*, and *measure*. You investigated the Main Menu and the Data Editor Toolbar and the options available for each of these. In the next chapter, you will learn how to save your data and output and how to get data and information in and out of SPSS.

Chapter 3

Getting Data In and Out of SPSS

△ 3.1 Introduction and Objectives

It is important that you save your data and output often in case your computer dies or the application you are using quits for no apparent reason. By "often" we do not mean every hour or so. We mean every few minutes. There may be occasions when you need to export your SPSS files to another application. Or you may need to import some files from other applications into SPSS. In addition, there are some handy SPSS sample files that were installed when SPSS was installed in your computer that many users do not know exist. We will be requesting that you use some of these sample files in certain exercises described in this book.

OBJECTIVES

After completing this chapter, you will be able to

Save and open your data and output files

Open and use sample files

Import files from other applications into SPSS

Export files from SPSS to other applications

Copy and paste data from SPSS to other applications

3.2 Typing Data Using the Computer Keyboard △

A simple method of entering data and other information into SPSS is to type it in using the computer keyboard. Whether you are a proficient typist, or you use the "hunt and peck" or "peer and poke" method, your information must be entered into cells in the Variable View and Data View screens. We use a balloon to indicate in which cell you are to type the first piece of information, as shown in Figure 3.1.

Figure 3.1 Use of the Balloon

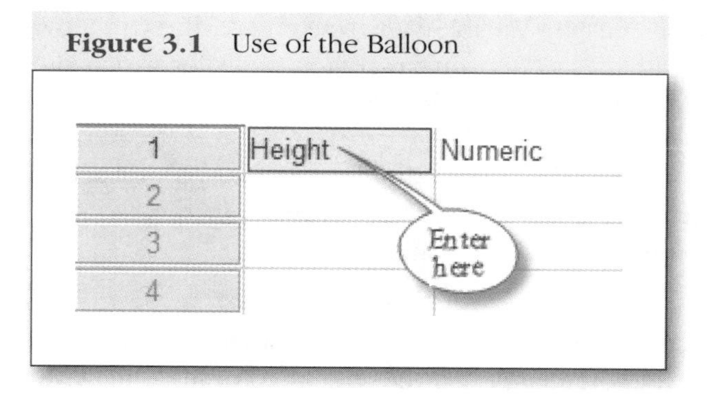

3.3 Saving Your SPSS Data and Output Files △

It is important that you understand the following convention of SPSS. When you save either a Data View screen or a Variable View screen, both screens are saved together when you click **Save As** on the Main Menu. To save these screens,

- Click **File** on the Main menu, and then click **Save As**. A window titled "Save Data As" will open. A folder titled "Documents" will normally appear in the "Look in" box indicating the SPSS default folder in which your files will be saved. Type a name for the file you are saving in the "File name" box.
- Click **Save**.

If you have requested an analysis or graph related to the data in the Data View screen, that output is displayed in a window titled "Output SPSS Statistics Viewer" that floats over the Data View screen. To save the Output-Viewer that contains your statistical tables and graphs, you must, in the Output-Viewer,

- Click **Save As**. A window titled "Save Output As" will appear, and you must indicate the location where you wish the file to be saved. The folder titled "Documents," which is the SPSS default folder, normally appears in the "Look in" box. Type a name for the file you are saving in the "File name" box.
- Click **Save**.

Important

The following is a simple procedure to use to ensure that you have saved the Data View screen, the Variable View screen, and the Output-Viewer screen associated with these. After you have entered all information into the Variable View screen and all data into the Data View screen,

- Click **Save As**. A window titled "Save Data As" will appear. The folder titled "Documents" normally appears in the "Look in" box. Type **chapter1** in the "File name" box.
- Click **Save**.

After saving these screens, request an analysis of the data and any graphs. After reviewing the Output-Viewer screen, you can either save or delete it. Since you have already saved your Data View screen and your Variable View screen, you can always open these anytime and again perform any analysis you wish and request any graphs you wish.

When you save a file, SPSS automatically places an extension at the end of the name you give to that file. If the file is a Data View or Variable View file, the extension is .sav, and the icon looks like the one in Figure 3.2. If the file is an output file, the extension is .spv and the icon looks like the one in Figure 3.3.

Figure 3.2 Data View Icon

Figure 3.3 Output Icon

It is important to realize that you can save a Data View screen or Variable View screen with the same name as an Output-Viewer screen because the file extensions are different. Consequently, one will not overwrite the other.

3.4 Opening Your Saved SPSS Files △

Opening your existing SPSS files is easy. To open a data file, from the Main menu,

- Click **File**, select **Open**, and then click **Data**. A window titled "Open Data" will appear requesting that you locate the file you wish to open. The word "Documents" will normally appear in this window as the SPSS default folder that contains the file you wish to open.
- Click the file name you wish to open.
- Click **Open**. If this is not the location of your file, click the down arrow in the "Look in" box and scroll to locate your file.
- Click the file name, and it will appear in the "File name" box.
- Click **Open**.

If you are opening an Output file, click **File**, select **Open**, and click **Output**. Then follow the steps for opening a data file.

For example, if you wish to open the data file "chapter1" that you saved when reading the first chapter,

- Click **File**, select **Open**, and then click **Data**. The folder titled "Documents" will appear in the "Look in" box.
- Click **chapter1**, and then click **Open**.

If you wish to open the Output file that you saved,

- Click **File**, select **Open**, and then click **Output**.
- Click **chapter1**.
- Click **Open**.

3.5 Opening SPSS Sample Files △

In this book, we refer to a group of data that are to be entered in Data View as a Data Set. Certain exercises in this book request that you enter data from the computer keyboard. These Data Sets are listed in Appendix A. When you are in Data View, simply turn to the appropriate section of Appendix A and enter the data listed on that page.

To save you time and toil in entering larger sets of data that may be required to satisfy, for example, assumptions regarding certain statistical tests, we will request that you open data files on your computer that contain this information. When SPSS was installed in the computer you are using, many useful SPSS sample files were included in that installation. These are some of the files we will request that you open. Follow these steps to access these files when requested in the book.

If you are using a PC, follow these steps:

- Start SPSS if it is not already running.
- On the SPSS Menu, click **File**, select **Open**, and then click **Data**.
- In the "Open data window," click the down arrow in the "Look in" box and scroll to and click the **C drive**.
- When a window opens, double click **Program Files**.
- When a window opens, scroll to SPSSInc. Double click **SPSSInc**.
- When a window opens, double click **SPSS Statistics 18**.
- When a window opens, double click **Samples**.
- When a window opens, double click **English**.
- A window will open in which you will see a long list of files you can open.

If you are using a Macintosh, follow these steps:

- Start SPSS if it is not already running.
- On the SPSS Menu click **File**, select **Open**, and then click **Data**.
- When a window opens, click **Applications**.
- In a side window, scroll to SPSSInc and click **SPSSInc**.
- When a side window opens, click **SPSS 18** or whatever version number shows.
- When a side window opens, click **Samples**.
- In a side window, you will see a long list of files you can open.

△ 3.6 Copying and Pasting Data to Other Applications

A simple method to transfer data from SPSS to other applications such as Excel, and to transfer data from these applications to SPSS, is the familiar copy and paste procedure. We will describe the transfer of data from the file, class_survey1, to Excel.

- Start SPSS.
- Locate and open **class_survey1**.

- Open Excel sheet 1.
- In SPSS, click on the first cell below exam1_pts and drag down to select all the 37 cells with data.
- Press **Ctrl+C**.
- In Excel, click the first cell below Column A.
- Press **Ctrl+V**, and the data from SPSS will be transferred to the 37 cells below Column A.
- In SPSS, click the first cell below exam2_pts and drag down to select all 37 cells with data.
- Press **Ctrl+C**.
- In Excel, click the first cell below Column B.
- Press **Ctrl+V**, and the data will be transferred to all 37 cells below Column B.

To copy data from Excel to SPSS, simply reverse the procedure.

3.7 IMPORTING FILES FROM OTHER APPLICATIONS

Excel is a popular and readily available spreadsheet, so we will use it as an example of an application from which you may wish to import data files into SPSS.

- On the Main Menu, click **File**, select **Open**, and then click **Data**. A window titled "Open Data" will appear. In the "Files of type" box, use the arrow to scroll and then select the file type **Excel**. The file you wish to open should be in the "Documents" folder.
- Click the Excel file name.
- Click **Open**. A window titled "Opening Excel Data Source" will appear.
- Click **OK**, and the file will be opened in the SPSS Data View screen.

3.8 EXPORTING SPSS FILES TO OTHER APPLICATIONS

An easy procedure to export SPSS files for use with other applications such as Excel is to use the "Save As" command from the Main Menu.

- With the file you wish to export open in SPSS, click **File** and then click **Save As**. A window titled "Save Data As" will open. Type a name for the file you wish to export in the "File name" box. Choose a destination for the file by clicking the arrow in the "Look in" box, which should be displaying the name "Documents." If this is not the correct folder, click the arrow and scroll to the correct folder.

- Click the arrow in the "Save as type" box, and scroll to the name of the application that will be used to open the file you are saving. There may be different versions of the same application listed. You should choose the version that will most likely be used to open the file. For example, there are several selections for Excel. Not all users of Excel will be able to use the newest version listed, depending on the version of Excel they have installed in their computer. Consequently, it may be wise to choose an earlier version even if that version may lack some of the bells and whistles of the newer version.
- So click **Excel 97 through 2003**, and then click **Save**. Note that the file extension for Excel is .xls. Your file has now been saved as an Excel file.
- If you start Excel on your computer, you will be able to click **Open** and load this file.

△ 3.9 Summary

In this chapter, you learned how to save and open your data and output files. You learned how to open the sample files that were installed with SPSS. You also learned how to import files from other applications into SPSS and how to export files from SPSS to other applications. In the next chapter, we will describe levels of measurement, including nominal, ordinal, and interval, and the necessity of carefully describing your data using these levels of measurement.

Chapter 4

Levels of Measurement

4.1 Introduction and Objectives △

Thus far, we have led you through the basic procedures needed to enter variables and data and how to navigate in SPSS. Chapter 4, on levels of measurement, covers an essential bit of knowledge required to successfully use SPSS—specifying the correct level of measurement for each of your variables. SPSS provides the user with three choices when selecting these levels of measurement: (1) nominal, (2) ordinal, and (3) scale. The major purpose of the chapter is to assist you in selecting the appropriate level of measurement for each of your variables.

Each of the three levels provides the analyst with different amounts of analyzable information. Another important consideration is that different statistical procedures require that the data be collected at specific levels of measurement. The level of measurement is partially determined by the basic nature of the variable (more on this later); however, the analyst does have a certain degree of freedom when specifying one of the three levels.

By way of definition, we say that *level of measurement* is a phrase that describes how measurable information was obtained while observing variables. For example, measuring an object with a ruler yields scale data; ranking first-, second-, and third-place winners provides ordinal data; and finally, when the females in a room are counted, we have nominal data. The information presented in the following sections is concerned with the expansion of this basic definition of levels of measurement.

OBJECTIVES

After completing this chapter, you will be able to

- Describe the characteristics of nominal, ordinal, and scale levels of measurement
- Distinguish between nominal, ordinal, and scale levels of measurement
- Determine the level of measurement appropriate for a variable

△ 4.2 Variable View Screen: Measure Column

By this time you should feel confident in navigating the various menus and windows. You have used the Variable View screen when you named variables. You may have noticed that one column on the Variable View screen was called *Measure*. The function of the *Measure* column must be addressed right away. To help you understand the concept of levels of measurement, and the proper use of the *Measure* column in SPSS, open the data file you saved in Chapter 1 that you titled as **chapter1**.

- Start SPSS, and then Click **Cancel** in the SPSS Statistics opening window.
- Click **File**, select **Open**, and then click **Data**.
- Click **chapter1.sav**, and then click **Open**.
- Click **Variable View** (at the bottom of the screen), and then inspect the *Measure* column that shows "scale."
- Click the cell below the *Measure* column (you will see a drop-down menu displaying, scale, ordinal, and nominal, as shown in Figure 4.1).

Figure 4.1 Variable View Screen: Measure Column Drop-Down Menu

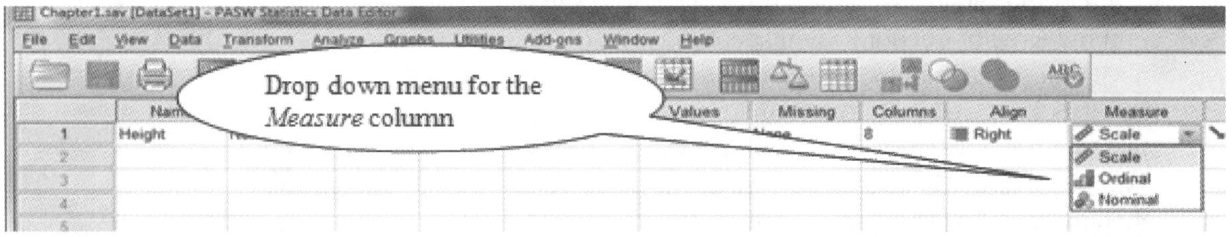

These three choices, as shown in Figure 4.1, are known as levels of measurement, which is the subject of this chapter.

SPSS, by default, selects "scale" for all numerical data that you enter in the Data View screen. SPSS does not always make the correct decision when it designates data as having been measured at the scale level. The SPSS program cannot consistently recognize the correct level of measurement that was used to measure your variables. Because of this, you, the person trying to make sense out of the data, must understand whether the attributes of variables were measured at the nominal, ordinal, or scale levels. Once you determine the correct level of measurement for the attributes of your variables, you then specify this in the *Measure* column of the Variable View screen. We are going to spend time helping you understand how to do this for each of your variables. We will give you just enough information needed to ensure that SPSS will correctly understand your data.

SPSS recognizes three levels of measurement that should be considered as a hierarchy: nominal (lowest level), then ordinal (middle level), and finally scale (highest level). This hierarchy (lowest, middle, and highest) refers to the amount of analyzable information contained in the numbers (your data) resulting from your counting or measurement of your observations. Let's look at a few examples that demonstrate these levels of measurement, beginning with the lowest level, nominal.

4.3 Variables Measured at the Nominal Level △

When a variable is said to have been measured at the nominal level, the numerical values only represent labels for various categories of that variable. It should be noted that when a variable is measured at the nominal level, the categories are often referred to as attributes of that variable. It is a common practice to refer to variables as being "measured" at the nominal level, but in reality we are only counting the occurrences of the variable's attributes. The numbers used to represent various categories (attributes) contain the least amount of analyzable information of the three levels of measurement. We can only count and calculate the percentage of the total number of observations that occupy the various categories. When trying to make sense of the attributes of variables measured at the nominal level, we may use data tables and bar graphs to display the numbers and/or percentages of the various attributes that were observed. Additional analytic methods, both descriptive and inferential, for the attributes of variables measured at the nominal level are described in future chapters.

Variables having separate categories that are not related in terms of quality or quantity are said to be measured at the nominal level. An example of a variable measured at the nominal level is gender. Gender has two possible categories: female, which could be labeled as Category 1, and male, labeled as Category 2.

You should note that the categories for gender, female and male, do not differ, at least in any meaningful way, in quality or quantity. In other words, the value of 2 that was assigned to males has no numerical meaning; the value 2 may be twice the value of 1, but males are not twice the value of females.

It is important to note that numerical values such as the labels 1 = *female* and 2 = *male* can only be counted (not added, subtracted, multiplied, or divided). However, understand that SPSS will incorrectly calculate statistics requiring such mathematical operations. For this and other reasons, you must understand levels of measurement when using SPSS.

Some additional examples of variables having attributes that could be measured at the nominal level are religious affiliation, name of automobile owned, city of residence, university attended, high school attended, neighborhood of residence, and voluntary membership in community organizations. It is important to note that the intention or purpose of the individual making the observations (collecting the data) is important since the attributes of variables in the above list could also be measured at the ordinal level. As an example of this dual measurement potential, let's look at university attended where intention could ultimately determine level of measurement. If the intention of the data collector is to infer that the quality of education depends on the university attended, then the numerical values representing the attributes (categories) take on the quality of ranked data, which is discussed in the next section. If this is not the purpose or intention of the data collection, then the attributes of the variable, university attended, take on the quality of the most basic level—nominal.

Another example of this dual measurement potential would be the name of automobile owned. During an application process, a prospective employer might ask the make and year of the automobile you currently own. The purpose of this question is to rank individuals on some quality—perhaps ambition. However, the more common interpretation of such a variable would be a simple counting of individuals owning various types (categories) of automobiles. It is important to understand the intention and purpose of your data collection before assigning a level of measurement to the attributes of variables.

△ 4.4 VARIABLES MEASURED AT THE ORDINAL LEVEL

Data measured at the ordinal level are often called ranked data because the categories of such variables measure the amount (a quality or quantity) of whatever is being observed. Thus, the categories of ordinal data progress in some systematic fashion; usually such variables are ranked from lowest to highest but might also be ranked from highest to lowest.

The ordinal level goes beyond the nominal in that it contains more analyzable information. The numerical values, which label the categories, are ordered or ranked in terms of the quality or quantity of the characteristic of interest that they possess. Think of the situation where your intention is to measure the degree of military rank possessed by a group of United States Marines. When collecting data you could specify the following: 1 = *private*, 2 = *corporal*, 3 = *sergeant*, and 4 = *general*. In this example, it can be seen clearly that as the numbers increase so does the degree of military rank. When using the ordinal level of measurement, it is important to note that the differences (distances) between ranks are undefined and/or meaningless. In this military rank example, the differences or distances between private and corporal are not the same as those between sergeant and general. The difference between a sergeant and general is many times greater than the difference between a private and corporal. If we input the numbers that were used to distinguish between these categories of military rank (1, 2, 3, and 4) into SPSS, we can do the same analysis that we did with attributes of variables measured at the nominal level. However, in addition to tables, graphs, and the calculation of percentages of each attribute, we can directly calculate the median. The median will then accurately portray the middle point of the distribution, which in this example would be the rank at which 50% of the individuals counted fall above and 50% fall below.

Another example of ordinal data is socioeconomic status, which could be described in terms of lower, middle, and upper class. We could specify that 1 = *lower class* (income of $0–20,000), 2 = *middle* ($20,001–100,000), and 3 = *upper* (>$100,000). Socioeconomic status is now categorized as 1, 2, or 3. Individuals so classified on their range of income are said to have been measured at the ordinal level.

One final example of a variable's attributes measured at the ordinal level would be a survey questionnaire seeking information about marital relationships. Suppose there was a question that sought to measure a person's satisfaction with his or her spouse. The variable might be titled "spousal satisfaction," and the attributes could be the following: 1 = *ready for divorce*, 2 = *thinking of divorce*, 3 = *satisfied*, 4 = *happy*, and 5 = *extremely happy*. The numerical data (1–5) could be entered into SPSS and analyzed using the mode and median for basic descriptive information.

4.5 VARIABLES MEASURED AT THE SCALE LEVEL △

Data measured at the scale level of measurement, as defined by SPSS, actually consists of data measured at two separate levels known as the interval and ratio levels. SPSS does not require that you, the user, distinguish between data measured at the interval and ratio levels. For mathematicians, the difference between the interval and ratio data is important, but for the types of applied research conducted when using SPSS, the differences are not important. Data measured at the SPSS's highest level, scale, contains more analyzable information than the nominal and ordinal levels because the differences or distances between any two adjacent units of measurement are assumed to be equal. This quality of "equal distances" permits all the mathematical operations conducted on the nominal and ordinal levels plus many more. When using the scale level of measurement, you may now summarize your data by computing the mean and various measures of dispersion such as the standard deviation and the variance. The mean and a measure of dispersion provide a distinct picture of the collected data—making it more understandable.

It should be noted that we drop the references to categories and/or attributes when using the scale level of measurement. We now refer to the "units of measurement" that are used to determine the numerical values associated with our variables. An example is measuring the distance to the top of a mountain. The "units of measurement" for the variable "distance" could be given in feet, megalithic yards, meters, kilometers, or any acceptable unit that might serve your purpose.

Examples of data that you could specify as scale data would be weight, height, speed, distance, temperature, scores on a test, and IQ. Let's look at height as a way of explaining the scale level of measurement. If you measure height in inches, you can state that it has equal intervals, meaning that the distance between 64 and 65 would be identical to the difference between 67 and 68. Or for temperature the difference between 72 and 73 is the same as between 20 and 21 degrees. Height also has a true (absolute) zero point; thus, the mathematician would say it is measured at the ratio level. For SPSS users, we only need to refer to it as data measured at the scale level.

Another example of a variable measured at the scale level is IQ. The difference between an IQ of 135 and 136 is considered the same as between 95 and 96. However, it should be understood that persons having an IQ of 160 cannot say they are twice as intelligent as persons with an IQ of 80, because IQ is measured at the interval, not ratio, level. Remember that variables measured at the ratio level require a true (absolute) zero point. It would be difficult, if not impossible, to show that an individual has zero intelligence.

△ 4.6 Summary

In this chapter, you had the opportunity to learn how to distinguish between variables measured at the nominal, ordinal, and scale levels of measurement. It is very important for you to understand that the level of measurement you select for your variable determines the type of statistical analysis you may use in analyzing the data. Certain types of analysis are only appropriate for specific levels. An example would be that you never calculate the mean and standard deviation for data measured at the nominal or ordinal levels. The computer and SPSS would do it, but it is an incorrect statistical analysis.

Once the appropriate level of measure has been inputted into SPSS, it is assigned an icon. These icons make the identification of levels of measurement for variables straightforward, thus facilitating the selection of the appropriate analytic techniques for various variables.

- The icon for variables measured at the nominal level is  .
- The icon for variables measured at the ordinal level is .
- The icon for variables measured at the scale level is .

In subsequent chapters, you will be given additional examples of how variables measured at the nominal, ordinal, and scale levels are entered into the SPSS Variable View and Data View screens. As you gain more experience, your confidence will grow in your ability to specify the correct level of measurement for your variables. The following chapter presents some basic validation procedures for variables measured at the nominal, ordinal, and scale levels.

Chapter 5

Entering Variables and Data and Validating Data

5.1 Introduction and Objectives

An important component of learning how to use SPSS is the proper and correct entry of variables and data and validating (verifying) these entries. Entering and assigning attributes (properties) such as *Measure* and *Values* to a variable is accomplished in the Variable View screen as described in Chapter 2. Entry of data is accomplished in the Data View screen. The Variable View screen and the Data View screen are known collectively as the Data Editor. Once variables and data are entered, SPSS can perform any number of statistical analyses and generate an extraordinary number of graphs of various types. However, if there are errors in either the data or the attributes assigned to variables, any analysis performed by SPSS will be flawed, and you may not even realize it. For example, if you assign scale in the *Measure* column to a nominal variable, SPSS will attempt to perform mathematical analysis on a nominal measure, and the results can be quite interesting.

In this chapter, you will enter the Data Set displayed in Appendix A (Tables A.1 and A.2), titled class_survey1. We will take you, step-by-step, through the process of correctly entering the variables, attributes, and data. If you do not assign attributes to each variable, SPSS will assign default values for some that may or may not be appropriate. For example, if you do not assign a level of measurement for a variable, SPSS will assign a level of measurement for that variable that may not be correct.

OBJECTIVES

After completing this chapter, you will be able to

Enter variables and their attributes in the Variable View screen

Enter data associated with each variable in the Data View screen

Validate the accuracy of the data

5.2 ENTERING VARIABLES AND ASSIGNING ATTRIBUTES (PROPERTIES)

The variables and data you are to enter into SPSS are displayed in Data Set 1 as shown in Appendix A in Tables A.1 and A.2. Table A.1 displays the seven variables and their attributes. Table A.2 displays the data associated with each of the seven variables. You will enter the variables and attributes in the Variable View screen and then enter the associated data for these variables in the Data View screen.

- Start SPSS and click **Type in data**.
- When a screen opens, Click **Variable View** at the bottom.
- Click the cell below *Name* and enter **Class**.
- Click the cell below *Type* and a window titled Variable Type will open. Click **Numeric** and then click **OK**.
- Click the cell below *Width* and then use the up-down arrows to select **8**.
- Click the cell below *Decimals* and use the up-down arrows to select **0**.
- Click the cell below *Label* and type **Morning or Afternoon Class**.
- Click the cell below *Values* and a window will open titled Value Labels. Type **1** in the Value box and type **Morning** in the Label box. Click **Add**. Then go back to the Value box and type **2** and in the Label box type **Afternoon**. Click **Add** and then click **OK**.

Your window should look like Figure 5.1.

Figure 5.1 Value Labels for Class

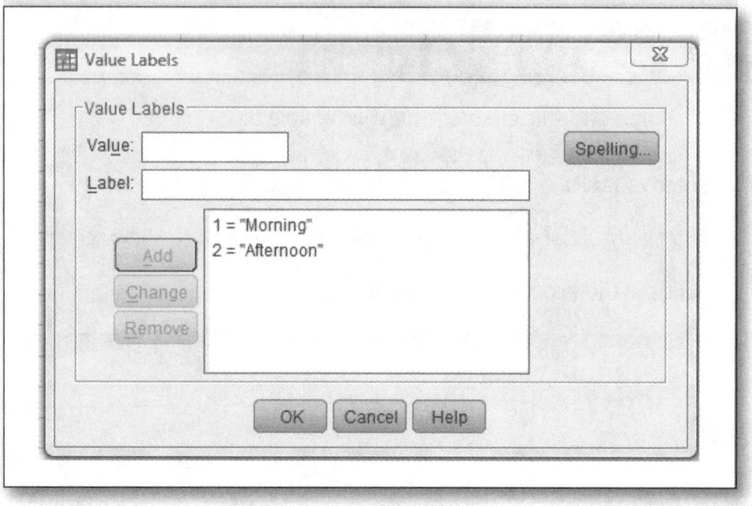

- Click the cell below *Missing* and a window titled Missing Values will open. Click **No missing values**, and then click **OK**.
- Click the cell below *Columns* and use the up-down arrows to select **8**.
- Click the cell below *Align* and use the up-down arrows to select **left**.
- Click the cell below *Measure* and use the up-down arrows to select **nominal**.

You have now entered all the attributes for the variable Class.

- Click the second cell below *Name* and enter **exam1_pts**.
- Click the second cell below *Type* and a window titled Variable Type will open. Click **Numeric** and then click **OK**.
- Click the second cell below *Width* and use the up-down arrows to select **8**.
- Click the second cell below *Decimals* and use the up-down arrows to select **0**.
- Click the second cell below *Label* and type **Points on Exam 1**.
- Skip the second cell below *Values* and click the second cell below *Missing*. A window will open titled Missing Values. Click **No missing values** and click **OK**.
- Click the second cell below *Columns*. Use the up-down arrows to select **8**.
- Click the second cell below *Align* and use the up-down arrows to select **Left**.
- Click the second cell below *Measure* and use the up-down arrows to select **Scale**.

You have now entered all attributes for the variable exam1_pts.

- Click the third cell below *Name* and enter **exam2_pts**. For this variable, enter exactly the same information you entered for exam1_pts, EXCEPT in the second cell below *Label*, type **Points on Exam 2**.

You have now entered all the attributes for the variable exam2_pts.

- Click the fourth cell below *Name* and enter **predict_grde**. All entries are the same EXCEPT *Label*, *Value*, and *Scale*. Click the fourth cell below *Label* and type **Student's Predicted Final Grade**.
- Click the fourth cell below *Values*, and a window will open titled Value Labels. Type **1** in the Value box and type **A** in the Label box. Click **Add**. Then go back to the Value box and type **2** and in the Label box type **B**. Click **Add**. Repeat for **3** = **C**, **4** = **D**, and **5** = **F**. Then click **OK**. Your window should look like Figure 5.2.

Figure 5.2 Value Labels for Predicted Final Grade

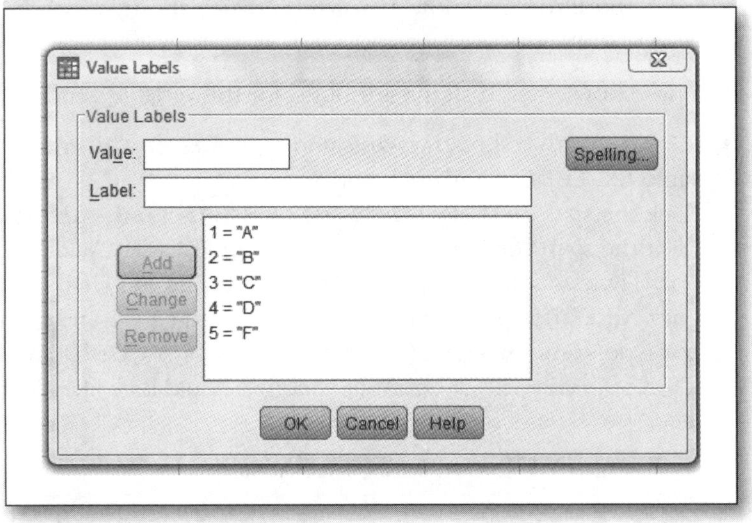

- Click the fourth cell below *Measure* and use the up-down arrows to select **Nominal**.

You have now entered the attributes for the variable predict_grde.

- Click the fifth cell below *Name* and type **Gender**. All entries are the same EXCEPT *Label*, *Values*, and *Scale*.
- Click the fifth cell below *Label* and type **Gender**.
- Click the fifth cell below *Values* and a window titled Value Labels will open. Type **1** in the Value box and type **Male** in the Label box. Click **Add.** Then go back to the Value box and type **2** and type **Female** in the Label box. Click **Add**. Click **OK**.

Your window should look like Figure 5.3.

Figure 5.3 Value Labels for Gender

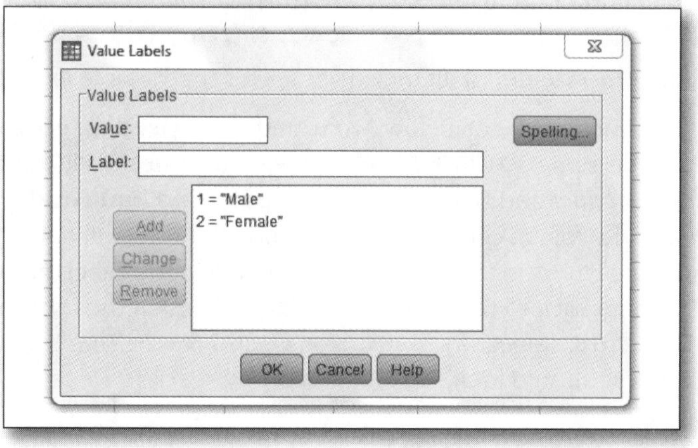

- Click the fifth cell below *Measure* and use the up-down arrows to select **Nominal**.

You have now entered all the attributes for the variable gender.

- Click the sixth cell below *Name* and type **Anxiety**. All entries are the same EXCEPT *Label*, *Values*, and *Scale*.
- Click the sixth cell below *Label* and type **Self-rated Anxiety Level**.
- Click the sixth cell below *Values* and a window titled Value Labels will open. Type **1** in the Value box and type **Much Anxiety** in the Label box. Click **Add**. Then go back to the Value box and type **2** and in the Label box type **Some Anxiety**. Click **Add**. Repeat for **3** = **Little Anxiety** and **4** = **No Anxiety**. Click **OK**. Your window should look like Figure 5.4.

Figure 5.4 Value Labels for Anxiety

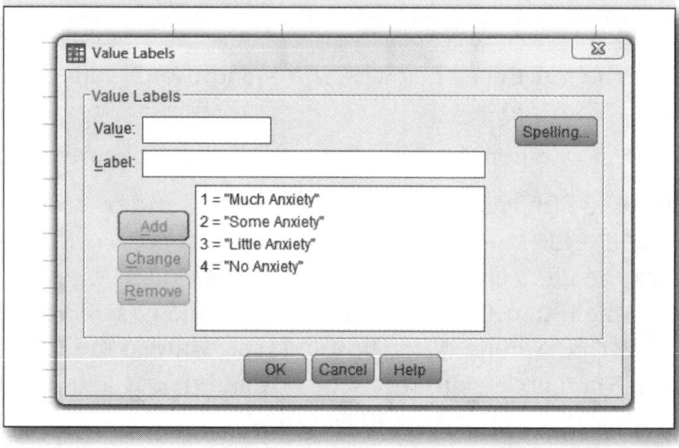

- Click the sixth cell below *Measure* and use the up-down arrows to select **Ordinal**.

You have now entered the attributes for the variable anxiety.

- Click the seventh cell below *Name* and type **rate_inst**. All entries are the same EXCEPT *Label*, *Values*, and *Measure*.
- Click the seventh cell below *Label* and type **Instructor Rating**.
- Click the seventh cell below *Values* and a window titled Value Label will open. Type **1** in the Value box and type **Excellent** in the Label box. Click **Add**. Then go back and type **2** in the Value box and type **Very Good** in the Label box. Click **Add**. Repeat for **3** = **Average**, **4** = **Below Average**, and **5** = **Poor**. Click **OK**.

Your window should look like Figure 5.5.

Figure 5.5 Value Labels for Instructor Rating

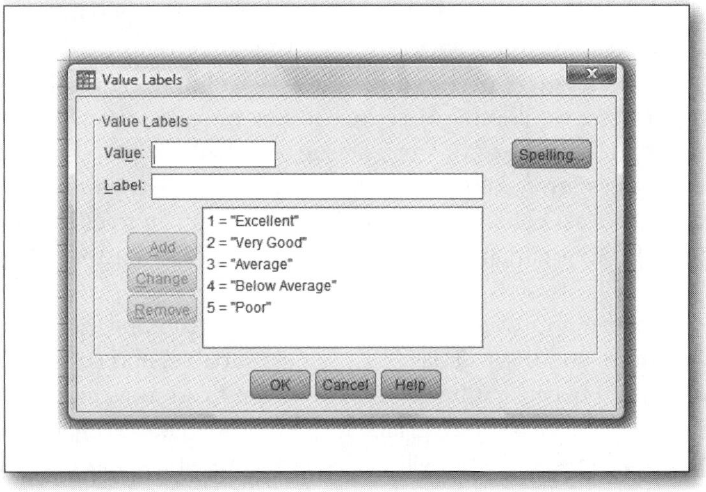

- Click the seventh cell below *Measure* and use the up-down arrows to select **Ordinal**.

You have now entered all seven variables and their attributes.

Figure 5.6 shows what your Variable View screen should look like after entering all the variables and attributes. If the variables and attributes in your screen do not match, now is the time to make corrections.

Figure 5.6 Variable View Screen for Class

	Name	Type	Width	Decimals	Label	Values	Missing	Columns	Align	Measure	Role
1	class	Numeric	8	0	Morning or Afternoon ...	{1, Morning}...	None	8	Left	Nominal	Input
2	exam1_pts	Numeric	8	0	Points on Exam One	None	None	8	Left	Scale	Input
3	exam2_pts	Numeric	8	0	Points on Exam Two	None	None	8	Left	Scale	Input
4	predict_grde	Numeric	8	0	Student's Predicted Fi...	{1, A}...	None	8	Left	Nominal	Input
5	gender	Numeric	8	0	Gender	{1, Male}...	None	8	Left	Nominal	Input
6	anxiety	Numeric	8	0	Self -rated Anxiety Level	{1, Much An...	None	8	Left	Ordinal	Input
7	rate_inst	Numeric	8	0	Instructor Rating	{1, Excellen...	None	8	Left	Ordinal	Input

△ 5.3 ENTERING DATA FOR EACH VARIABLE

At this point, you will enter data from Table A.2 in Appendix A for each variable in what may seem a somewhat tedious process. Although tedious, it is extremely important that you accurately enter the data and verify your entries by visual inspection.

Click **Data View** at the bottom of the screen. The variables you have entered are displayed across the top of the screen. You will enter data, starting in the first cell (intersection of Row 1, Column 1) below class, from left to right across columns. Use the Tab key or right Arrow key to move from one cell to another. We strongly suggest that you place a ruler under the first row of data to block out the other data. When you have entered the data for the last cell in a row, use the mouse button to click the second cell under the first variable, class, and enter data for that row. Repeat for all data.

An alternative approach is to enter all data for the first variable moving down a column, and then all data for the next variable, and so on. We prefer the first method because it involves less eye and head movement.

Before quitting SPSS, be certain to save your data using the filename **class_survey1**, because you will be using the variables and data in subsequent chapters.

△ 5.4 VALIDATING DATA

For smaller data sets with fewer than 50 cases and many variables, it is usually sufficient to validate the entries by visual inspection. But when there are hundreds or thousands of cases, and/or many variables, validation by inspection to determine if each case has been correctly entered can be an overwhelming task. Fortunately, there are other methods such as using the Validate Data module. But this module is not included in the base SPSS installation and must be purchased and downloaded. On the Main menu, click **Add-ons**, select **SPSS Data Validation**, and click **Validate Data**. Your computer browser will take you to a website that explains the process of validating data. There is a module titled *IBM SPSS Data Preparation* available for download that includes the Validate Data procedure, among others. The Validate Data procedure enables you to apply rules to perform data checks based on each variable's measure level. You can check for invalid cases and receive summaries regarding rule violations and the number of cases affected. In general, the validation process involves applying a set of rules to a data set and asking SPSS to determine if those rules have been violated. The Validate module is very useful, but, unfortunately, it is not free.

We next describe some additional methods for data validation. In the following paragraphs, you will add fictitious data to the database just created and saved as **class_survey1** to demonstrate basic validation procedures. The most direct method to check for data entry errors is to produce frequency tables for your variables that were measured at the nominal or ordinal levels. For those variables measured at the scale level, a descriptive statistics table is useful. This approach to validation also permits a quick inspection of the variable labels and the values assigned to each of the categories of your variables.

The hands-on experience will involve the use of the data from the Class Survey used in previous sections. If the data file is not already open, open it now. Click **Data View** screen and enter the erroneous data in Rows 38 and 40 (no data in Row 39) as shown in Figure 5.7. The data entered into these rows will be used to illustrate the validation procedures for this new database consisting of 40 students.

Figure 5.7 Class Survey Data View Screen With Three Additional Cases

	class	exam1_pts	exam2_pts	predic_grde	gender	anxiety	rate_instrc
37	2	41	72	4	2	1	1
38	3	105	102	5	3	5	5
39	.	.	.	.	.	.	.
40	3	101	107	1	3	5	0

Once the new data are entered, we begin the validation process by looking at the variables measured at the nominal and ordinal levels. **Caution:** Do not save this new database without assigning it a new name!

Validation of Nominal and Ordinal Data

- Click **Analyze**, select **Descriptive Statistics**, and click **Frequencies**.
- Click, while holding down the **Ctrl** key, **Morning or Afternoon Class**, **Student's Predicted Grade**, **Gender**, **Self-rated Anxiety Level**, and **Instructor Rating** (variables measured at the nominal and ordinal levels).
- Click the right arrow.
- Click **OK**.

Once you have clicked **OK**, SPSS will produce six tables for your five variables. The first table is as shown in Figure 5.8. This table presents a summary of the five variables that you selected for this data check operation. There are five columns, each containing one variable and the number of values recorded for each.

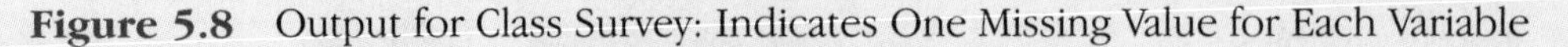

Figure 5.8 Output for Class Survey: Indicates One Missing Value for Each Variable

Statistics

		Morning or Afternoon Class	Student's Predicted Final Grade			Instructor Rating
N	Valid	39	[illegible]	39	39	39
	Missing	1	1	1	1	1

In the first row, you will see the labels of the variables you selected for the analysis. The next row is titled "N Valid," which only indicates that 39 numerical values were entered. It should not be inferred that these values were somehow correct. The next row, titled "Missing," informs you that there is one missing numerical value for each of the five variables. An inspection of the Data View screen reveals that the values for an entire case were omitted.

The next step in our data validation procedure is to determine if there are additional data entry problems by looking at the frequency tables for each of our variables. The first table, titled "Morning or Afternoon Class," is given in Figure 5.9.

Figure 5.9 Frequency Table: Output for Class Survey

Morning or Afternoon Class

		Frequency	Pe[illegible]	[illegible]	Cumulative Percent
Valid	morning	18	[illegible]	[illegible]	46.2
	afternoon	19	47.5	48.7	94.9
	3	2	5.0	5.1	100.0
	Total	39	97.5	100.0	
Missing	System	1	2.5		
Total		40	100.0		

Indicates entry error

First, look at the title of the table, which is the "Label" that was entered in the Variable View screen. Make sure the spelling is correct and the title accurately describes the variable. Look at the rows; in the first column, you should see "Morning," and in the Frequency column, "18." The titles for the rows, such as Morning, are actually the value labels you attached to the categories of your variables. The error in data entry can be easily spotted because of the 3 that is printed, as a title, for a row of data. It indicates that the incorrect value of 3 was entered two times. Since you had not assigned any value label to this incorrect entry, SPSS simply counted the number of times "3" was observed and then reported this in the frequency table. For this variable, there were only two possible categories: The value label for the Morning class was 1, whereas the value label for the Afternoon class was 2. Thus, all the entered values should have been 1 or 2. For small databases, it is then a simple matter to visually inspect the data file, locate the error, and correct the mistake. For larger databases, one should click the toolbar icon (called Find) shown in Figure 5.10. An alternate method used to activate the "Find" function is to click **Edit** on the Main Menu and select the same icon from the drop-down menu.

Figure 5.10 Find Icon

Clicking this icon opens the "Find and Replace—Variable View" window as shown in Figure 5.11. Type **3**, the incorrect entry identified in the frequency table, in the **Find** box and click **Find Next**. This procedure will then return you to the data view screen, and the incorrect value will be highlighted. The frequency table identified multiple incorrect entries of 3, so you simply click **Find Next** and SPSS will progress through the database stopping at each incorrect value.

Figure 5.11 "Find and Replace—Variable View" Window

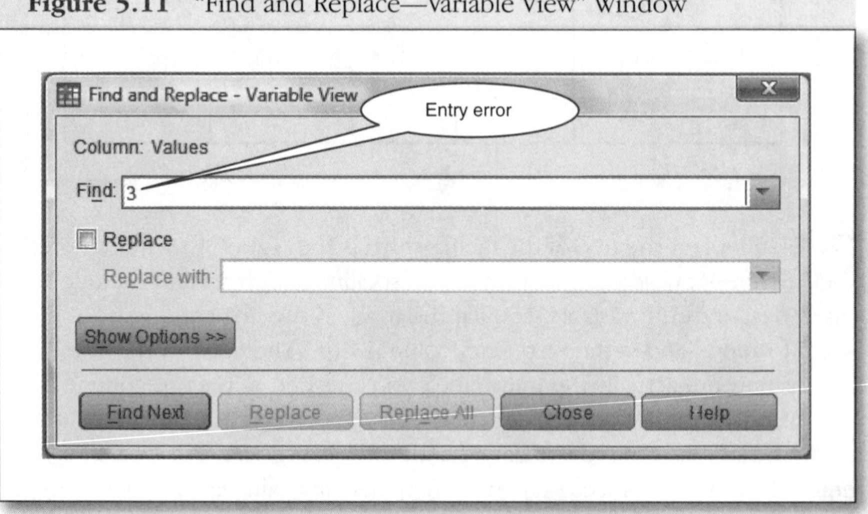

At this point, you should inspect the remaining four frequency tables in the output viewer and identify additional data entry errors for the remaining four variables.

Validation of Scale Data

To look for data entry errors for variables measured at the scale level, we will use the data entered for exam1_pts and exam2_pts. To do this validation procedure, we check the data to see if there are any values that are less than or greater than some expected value. For exam1 and exam2, we know that it was impossible to score less than 0 or more than 100 points; therefore, any values detected outside of this range would be incorrect. The following procedure will check the entered data for any values outside of the expected range.

- Click **Analyze**, select **Descriptive Statistics**, and click **Descriptives**.
- Click **Points on Exam One** and **Points on Exam Two** (variables measured at the scale level) while holding down the computer's **Ctrl** key.
- Click the arrow.
- Click the **Options** box (the "Descriptives: Options" window opens; see Figure 5.12).
- Unclick all boxes that may be checked and click **Minimum** and **Maximum**.
- Click **Continue**, and then click **OK**.

Your window should look like Figure 5.12.

Figure 5.12 "Descriptives: Options" Window

Once you click **OK**, a table titled "Descriptive Statistics" appears in the output viewer as shown in Figure 5.13. Inspection of this table easily identifies any value that may be less than or greater than any expected value. In this example, we know that the total possible points on these exams are 100; thus, we can easily spot the data entry errors of 105 and 107. In the "N" column, you also see that there is a missing value for both variables because N = 39, and it should be 40.

Figure 5.13 Output Viewer: Descriptive Statistics

Descriptive Statistics

	N	Minimum	Maximum
Points on 1st Exam	39	14	105
Points on 2nd Exam	39	23	107
Valid N (listwise)	39		

△ 5.5 SUMMARY

In this chapter, you learned how to enter variables and their attributes. You also learned how to enter data for each variable. In the next chapter, you will learn how to work with variables and data that have already been entered into SPSS.

Chapter 6

Working With Data and Variables

6.1 Introduction and Objectives △

Sometimes it is desirable to compute new variables and/or recode variables from an existing database. When computing a new variable, you often add a number of variables together and calculate an average, which then becomes the new variable. Recoding may involve the changing of a variable measured at the scale level into a variable measured at the nominal or ordinal levels. In this chapter, you will recode a variable measured at the scale level into a string variable. You may wish to do such operations to make the data more understandable or perhaps for more specific descriptive or inferential analysis. In this chapter, you will use various SPSS commands to compute and recode data, which results in the creation of a new variable.

The SPSS user may sometimes find it necessary to insert a case or variable into an existing database. These procedures are also explained, which provide many hands-on opportunities to use both the Data View and Variable View screens.

OBJECTIVES

After completing this chapter, you will be able to

Compute a new variable by averaging two existing variables

Recode a variable measured at the scale level into a new string variable

Insert a missing case (row of data) into a data set

Insert a variable (column) into an existing data set

Cut and paste existing data within the Data View page

△ 6.2 Computing a New Variable

For this exercise, you will use the **class_survey1** database created in Chapter 5. You will combine the first and second exam scores and create a new variable. The new variable will be the average of the scores earned on Exams 1 and 2.

- Start SPSS, and then click **Cancel** in the SPSS Statistics opening window.
- Click **File**, select **Open**, and then click **Data**.
- In the file list, locate and click **class_survey1.sav**.
- Click **Open**.
- Click **Transform** on the Main Menu, and then click **Compute Variable** (the Compute Variable window opens; see Figure 6.1).
- Type **avg_pts** in the Target Variable box (this is the name of your new variable as it will appear in the Data View screen).
- Click the **Type & Label** box and the "Compute Variable: Type and Label" window opens and floats in front of the Compute variable window (see Figure 6.2).
- In the **Label** box, type **Average Points On Exams** (this is your new variable's Label; see Figure 6.2).
- Click **Continue** (the window shown in Figure 6.2 disappears and Figure 6.1 remains).
- Click the parentheses button on the keypad (which will move the parenthesis to the "Numeric Expression" box).
- Click **Points on Exam One** (found in the variable list on the left side of the window), and then click the arrow to the right of the variable list.

Figure 6.1 Compute Variable Window: Database Named class_survey1

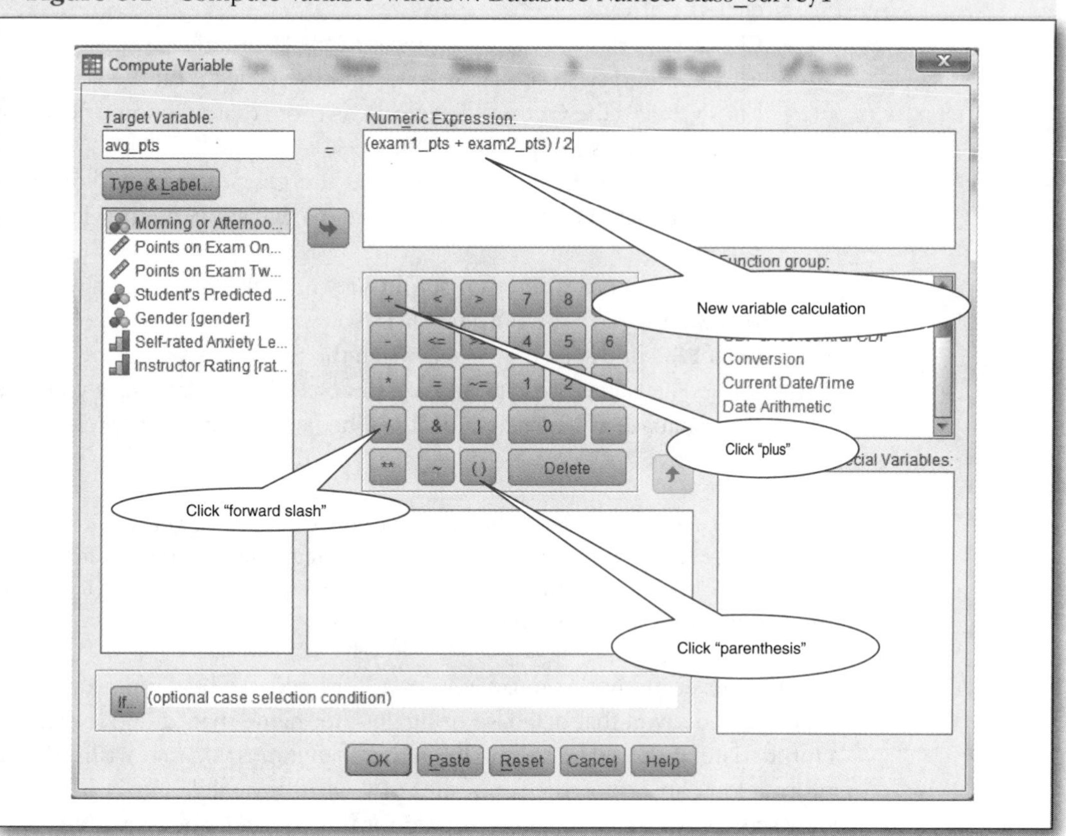

Figure 6.2 Compute Variable: Type and Label Window

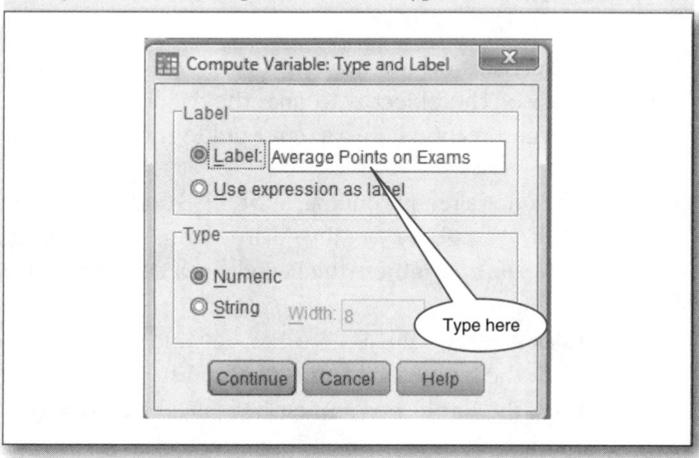

- Click + on the keypad, then click **Points on Exam Two**, and then click the arrow.
- Click to the right of the parenthesis in the "Numeric Expression box."
- Click forward slash on the keypad, and then click **2** on the keypad.
- Click **OK** and the Output Viewer opens (note that only some file information appears in the Output Viewer).
- Click the **x** in the red box to close the Output Viewer. Note that both the Data View and Variable screens will show the newly added variable.
- A window opens when the Output Viewer is closed that asks whether you want to save the output. Click **No**.
- Click **File**, and select **Save As**. When the Save Data As window opens, in the File Name box type **class_survey2**. (Note: Make sure you save this database as instructed, as it will be used in future chapters.)
- Click **Save**.

Note: If you feel comfortable with mathematical expressions, you can enter the expression **(exam1+exam2)/2** directly in the Numeric Expression box from your computer's keyboard. Note that the parentheses are required to force SPSS to add the two variables and then divide their sum by 2.

Whichever method you use, you have now added a new variable to your database and saved that database using another name that can be recalled at a future date. It should be kept in mind that sometimes you may wish to enter information concerning the properties of your new variable into the Variable View screen. No such action is required for this particular new variable.

△ 6.3 Recoding Scale Data Into a String Variable

For this exercise, you will use the variable (avg_pts) that was created in the prior exercise. The object is to take the average points (scale data) for the two exams and recode it into a letter grade (nominal data).

- If it's not already running, start SPSS and open **class_survey2.sav**.
- Click **Transform** on the Main Menu, and then click **Recode into Different Variables** (the Recode into Different Variable window opens; see Figure 6.3).
- Click **Average Points On Exams**, and then click the arrow.
- Click the **Name** box and type **Grade**.
- Click the **Label** box and type **Final Letter Grade**.

- Click **Change** (which moves the soon-to-be-created new Output Variable to the center panel, which now looks like: **avg_pts --> Grade**, and the Recode into Different Variables window now looks like Figure 6.3).

Figure 6.3 "Recode Into Different Variables" Window

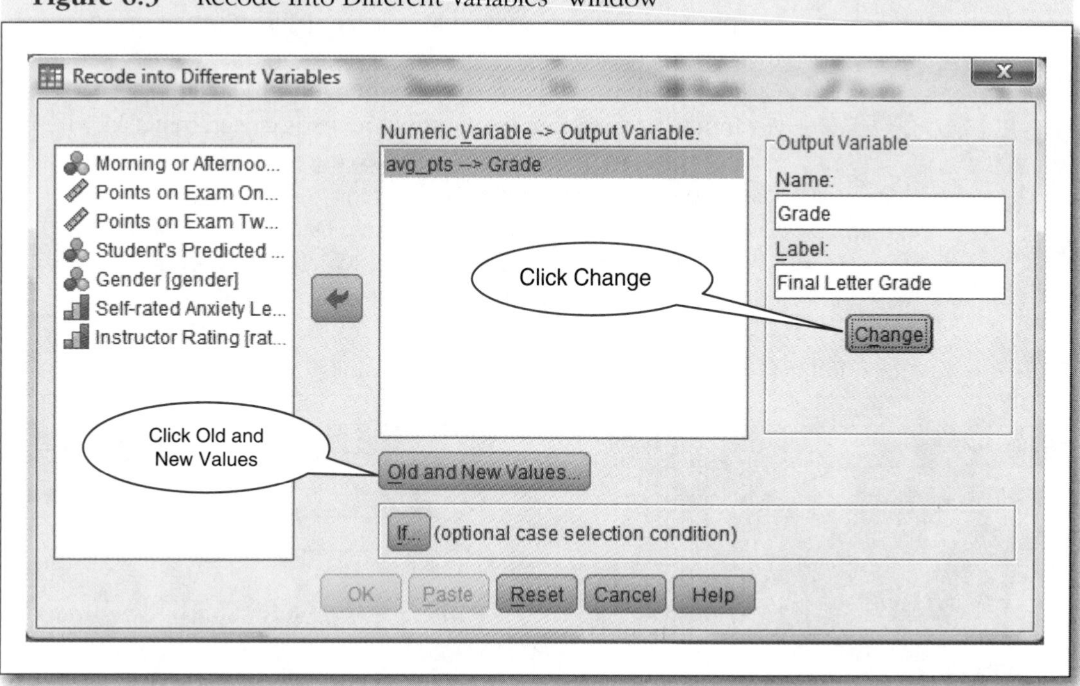

- Click **Old and New Values** (the Recode into Different Variables: Old and New Values window opens; see Figure 6.4).
- Click **Range**, and then type **0** in the upper box and **50** in the lower box.
- Click **Output variables are strings** (located on the lower right side of this window).
- Click **Value** and type uppercase **F** in the box; then click **Add**.
- Click the **Range** box, and then type **51** in the upper box and **69** in the lower box.
- Click **Value** and type uppercase **D** in the box; then click **Add**.
- Click the **Range** box, and then type **70** in the upper box and **79** in the lower box.
- Click **Value** and type uppercase **C** in the box; then click **Add**.
- Click the **Range** box, and then type **80** in the upper box and **89** in the lower box.
- Click **Value** and type uppercase **B** in the box; then click **Add**.

- Click the **Range** box, and then type **90** in the upper box and **100** in the lower box.
- Click **Value** and type uppercase **A** in the box (at this point, your window should look exactly like Figure 6.4 but without the balloons). Click **Add**.
- Click **Continue**, and then click **OK**.
- Output Viewer opens (note that only some file information appears in the Output Viewer).
- Click the **x** in the red box to close the Output Viewer.
- A window opens when the Output Viewer is closed that asks whether you want to save the output. Click **No** (the new variable appears in the Data View and Variable View windows).
- Click **File**, and then **Save As** (the Save Data As window opens).
- Click **class_survey2** in the data file, which moves **class_survey2** to the File Name box, and then click **Save**.

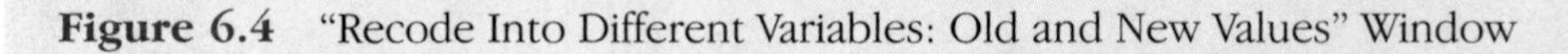

Figure 6.4 "Recode Into Different Variables: Old and New Values" Window

Once you click **Save**, a new database has overwritten the one you named in the prior exercise. This new database (**class_survey2**) now includes a scale variable called avg_pts and a string variable called Grade, which was created using the recode function available in SPSS. You will use this database in future chapters, so be sure to save it as instructed.

6.4 INSERTING NEW VARIABLES AND CASES INTO EXISTING DATABASES

You may sometimes find it necessary to enter a variable (a new column) or case (a new row) into an existing database. Perhaps you just forgot a case or variable or maybe you wish to update a database previously entered. SPSS provides two easy ways to accomplish this task. The first method is to click **Edit** on the Main Menu and then click **Insert Variable** or **Insert Cases** depending on your need. The second and more efficient method is to click on one of the icons found on the toolbar, which accomplishes the same task. Click the icon shown in Figure 6.5 when you need to insert a variable. Click the icon shown in Figure 6.6 if you must insert a case.

Figure 6.5 Insert Variable

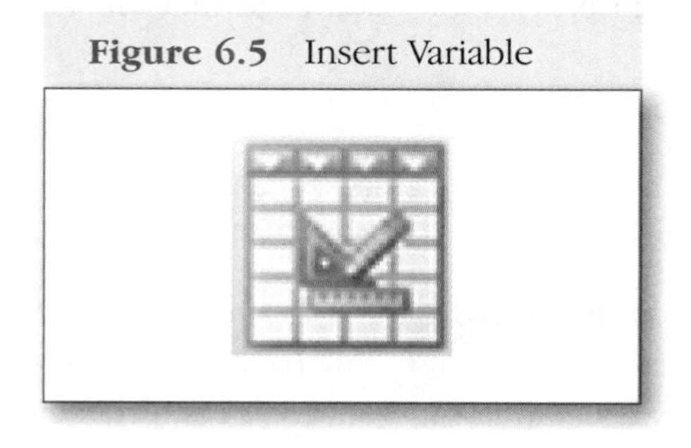

Figure 6.6 Insert Cases

To provide you with hands-on experience, open the database created in the prior section (**class_survey2.sav**) if it is not already running. Let's say you forgot to enter a variable that should have followed the variable called anxiety. If you click any column, the new variable is inserted to the left of that clicked column.

- Click **Data View** to open the screen.
- Click anywhere in the column to the right of anxiety.

- Click the toolbar **insert variable icon** in Figure 6.5 (a new column appears to the right of the anxiety variable column).
- You may now go to the Variable View screen and specify variable properties.
- Next you may go to the Data View screen and enter data for your new variable.

To insert a case (a row of data), a procedure similar to entering a variable is followed. Using the same database as in the prior section, let's assume you forgot to enter case number 10.

- In Data View click row **10**, then click the **insert cases icon** (Figure 6.6).
- A new row is created in row 10 that is ready to receive data.

It is recommended that you DO NOT save the changes made for the insert variable and insert case procedures as they will not be used in future chapters.

△ 6.5 DATA VIEW PAGE: COPY, CUT, AND PASTE PROCEDURES

Sometimes it is convenient to use the copy and paste function when entering or manipulating data on the Data View page. This is especially useful when entering repetitive codes for your variables. The procedure used by SPSS is identical to that used in all common word processing programs. We will demonstrate the method for PC users and then provide a brief explanation for the Mac user. We begin with the method for PC users.

- Start SPSS.
- Click **File**, select **Open**, and then click **Data**.
- Click **class_survey1.sav**, and then click **Open** (click **Data View** tab if the page is not open).
- Click the cell in Row 1, Column 1, and drag the mouse up to and including Case 18 (you now have a column of highlighted 1s; recall that this is the code used for males).
- Right click the mouse, and then click **Copy**.
- Move the mouse to Case 38 and right click the cell in Row 38, Column 1.
- Click **Paste**.

Once **Paste** is clicked in the above sequence, you will see the new "cases" added to your database. Let's use the "cut" function available in SPSS and return the database to its original version.

- Click the cell in Row 38, Column 1, and drag the mouse up to and including case 55.
- Right click the mouse, and then click **Cut** (do not save this altered database).

For the Mac computer user, the above procedure is approximately followed except that the edit function on the Main Menu is used. When clicking edit on the Main Menu, you are offered the "cut" and "paste" choices. We feel it is unnecessary to show bullet points, as any Mac user will understand the cut and paste method when using the edit menu.

Important Note: Once you have finished the above cut and paste exercises, make sure you DO NOT save the database. You will use the original **class_survey1** database many times in future chapters, and any inadvertent changes you may make will have a negative impact on your future work.

6.6 SUMMARY

In this chapter, you were given the opportunity to learn the basic techniques required to compute new variables from the values in an existing database. You first combined two variables that were measured at the scale level, found the average, and created a new variable. You then recoded this new variable into a letter grade to create a variable in string form. You also inserted new variables and new cases into an existing database.

In the next chapter, you will be given hands-on experience in using the various help features. Some of these help features are part of the SPSS program that was installed when you downloaded it onto your computer's hard drive. Other help features are based on links to online sites and as such must be used when your computer is online. All this will be explained in the following chapter.

CHAPTER 7

USING THE SPSS HELP MENU

△ 7.1 INTRODUCTION AND OBJECTIVES

SPSS offers a rich and extensive variety of features on the Help menu. You can obtain immediate assistance in understanding virtually any topic covered in SPSS. But the Help menu is not the only method to obtain assistance on particular topics. For example, when you request SPSS to compute a statistic or generate a graph, various windows will open. These windows contain options (choices), one of which is a Help option that you can click to obtain assistance concerning the statistics or graph you wish to generate.

There are 10 options available in the Help menu: Topics, Tutorial, Case Studies, Statistics Coach, Command Syntax Reference, SPSS Developer Central, Algorithms, SPSS Home, Check for Updates, and Product Registration. We will describe the first three options, and we will describe the purpose of the Help Search box. In addition, we will describe the Help available in windows that open when you request SPSS to perform a statistical analysis or generate a graph.

OBJECTIVES

After completing this chapter, you will be able to

- Describe the purpose of the following on the Help menu Topics: Tutorial, Case Studies
- Describe the purpose of the Help Search box
- Use the Topics option to locate and gather information on a topic of interest
- Use the Tutorial to select and run a tutorial concerning a major SPSS topic
- Use the Case Studies option to study and analyze a topic of interest
- Use the Search box to locate and study a topic of interest
- Use the Help selection available in Windows that open when selecting Analyze on the Main menu

7.2 HELP OPTIONS

Following is a brief description of the purpose of the first three selections in the Help menu:

Topics: Provides access to the Contents, Index, and Search tabs, which you can use to find specific Help topics. When you click **Topics** in the Help menu, your computer browser opens a window showing the topics available for your inspection. If you click **Contents**, a list of major topics will be presented from which you can choose. If you click **Index**, a panel containing the letters of the alphabet appears. When you click a letter, a listing of topics appears from which you can select.

Tutorial: Illustrated, step-by-step instructions on how to use many of the basic features in SPSS. You don't have to view the whole tutorial from start to finish. You can choose the topics you want to view, skip around and view topics in any order, and use the index or table of contents to find specific topics. When you click **Tutorial** in the Help menu, a window opens listing all the tutorial topics. Note: The tutorials provide descriptions and explanations for the various SPSS sample files that were installed on your computer, as discussed in Chapter 3 of this book.

Case Studies: Hands-on examples of how to create various types of statistical analyses and how to interpret the results. The sample data files used in the examples are also provided so that you can work through the examples to see exactly how the results were produced. You can choose the specific procedure(s) about which you wish to learn from the table of contents or search for relevant topics in the index.

△ 7.3 Using Help Topics

Let's get some assistance concerning the proper method of entering variable properties such as you entered in the exercise described in Chapter 5.

- Click **Help** on the Main menu, and then click **Topics**. A portion of the window that opens is displayed in Figure 7.1. In the top left section, you will see "Contents," "Index," and "Search."
- Click **Contents**. Double click **Data Preparation**, and then click **Variable Properties**. A window will open explaining how to define value labels, identify missing values, and assign levels of measurement such as nominal, ordinal, and scale levels.

Figure 7.1 Help Window

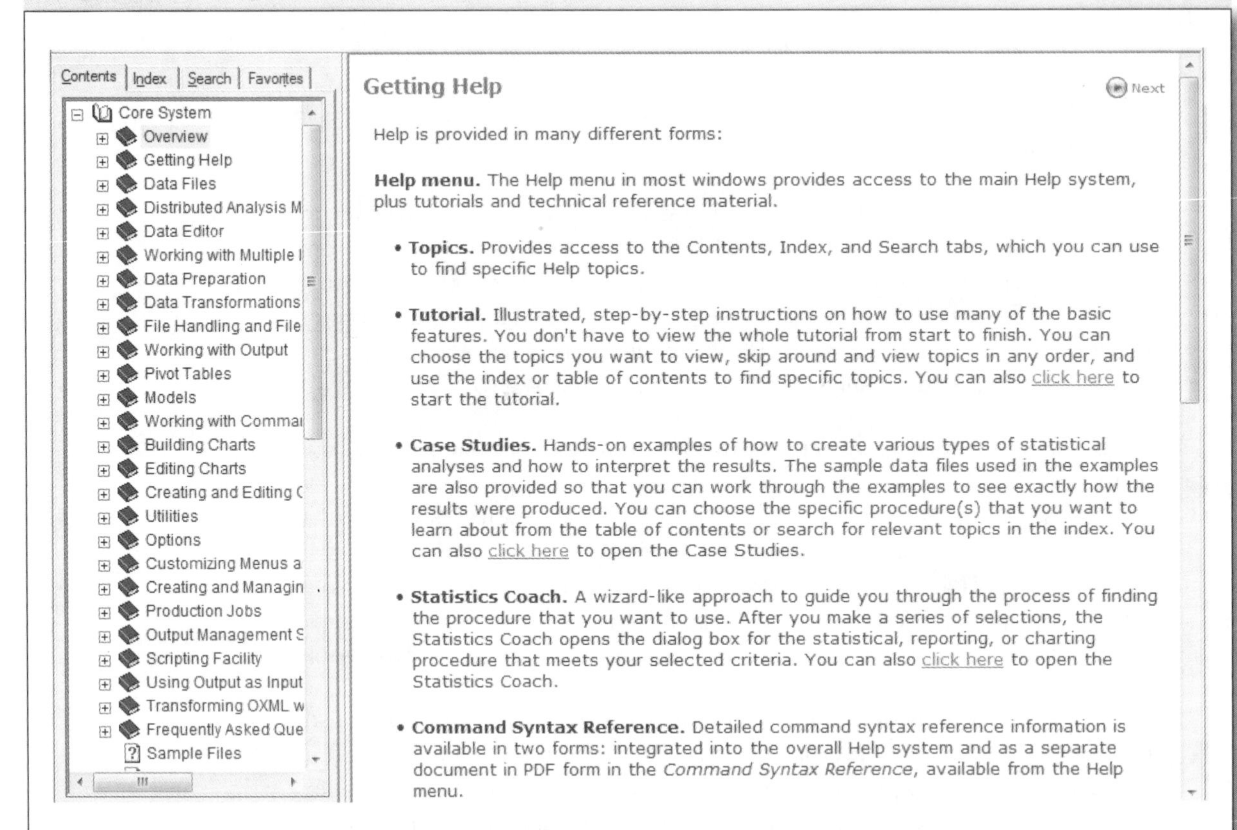

Let's investigate another topic.

- Click **Help** on the Main menu, and then click **Topics**. A window will open.
- Click **Contents**. Then double click **Core System**. Double click **Building Charts** and then click **Building a Chart from the Gallery**. A window will open that explains how to generate a graph (chart) using data you have collected. We will cover two of these methods of creating charts available in SPSS in Chapters 8 and 9. Note that in SPSS, the terms *graphs* and *charts* are used interchangeably.

Let's see what information we can find for the term *median* using Index.

- Click **Help** on the Main menu, and then click **Topics**.
- Click **Index**, and a panel will open displaying a very long list of topics in alphabetical order from which you may select. Use the scroll bar to find "median." A quicker way to find "median" is to type it in the "Keyword to find box."

Finally, let's see what information we can find concerning "measures of central tendency" using the Search option.

- Click **Help** on the Main menu, and then click **Topics**.
- Click **Search** and a search box appears. Type "measures of central tendency" in the box and click **List Topics**. Several options will appear. Double click any of these, and information will be displayed in a panel on the right.
- For example, double click **Frequency Statistics**, and you will be presented with information describing mean, median, and mode.

7.4 USING HELP TUTORIAL △

Let's run a tutorial that will explain how to enter variables and data into the Data Editor, assign variable labels, and much more. Note: The Variable View screen and the Data View screen are known collectively as the Data Editor.

- Click **Help** on the Main menu, and then click **Tutorial**. Click **Using the Data Editor**. A screen will open explaining the purpose of the Variable View screen, the Data View screen, and other items.
- On the bottom of the screen, click the right arrow that will take you to the next screen titled "Entering Numeric Data" in which you

will see a window showing the Data View screen, and on the right of that window, some comments regarding the process of entering data.

- Continue clicking the right arrow, and you will see additional screens with windows showing various information with accompanying explanations on the right side of the screen. If you continue to click that right arrow, you will learn how to accomplish tasks such as entering numeric and string data, adding variable labels, adding value labels, handling missing data, and much more. If you miss something, simply click the left arrow to return to the previous screen.

If you have questions regarding any topic in SPSS, it is likely you will find a tutorial to assist you in understanding. Some tutorials are rather lengthy and contain information you may have to spend some time reviewing to achieve a working knowledge. Tutorials require your patience and fortitude, but most are worth the effort.

△ 7.5 USING HELP CASE STUDIES

Using "case studies" provides you the hands-on opportunity to examine an actual database for a company, business, or other situation and generate statistics and graphs of interest to help you learn how SPSS can best assist you when, for example, you have collected your own data for analysis. Case studies use sample files that can be manipulated to answer questions. Essentially, one asks questions and then requests SPSS to provide answers, or at least output, that will help you answer these questions. The purpose and usefulness of using Case Studies is best illustrated with an example.

The following case study uses frequencies to study nominal, ordinal, and scale data. Imagine you manage a team that sells computer hardware to software development companies. At each company, your representatives have a primary contact. You have categorized these contacts by the department of the company in which they work (Development, Computer Services, Finance, Other, Don't Know). This information is collected in a sample file titled contacts.sav. (See Chapter 3, Section 3.5, Opening SPSS Sample Files.) You will use frequencies to study the distribution of departments to see if it meshes with your goals. In addition, you will study the distribution of company ranks and the distribution of purchases. Along the way, you will construct graphs and charts to summarize results.

- Click **Help** on the Main menu, and then click **Case Studies**. A window titled Table of Contents will open. Click **Statistics Base**, and then click **Summary Statistics Using Frequencies**.

- Click the right arrow at the bottom right of the screen to move from one screen to the next as you did in the tutorial exercise. The final screen shows the results of the investigation of data. You have assessed the distributions of the departments and company ranks of your contacts and the amounts of last sales. A major finding is that the distribution of sales is highly skewed as shown in the tables and graphs, indicating that a transformation of scales would be more appropriate in further analyses.

7.6 GETTING HELP WHEN USING ANALYZE ON THE MAIN MENU Δ

A convenient and easily accessible method of obtaining help is available when you choose a topic from Analyze on the Main menu. Make certain there are data in the Data Editor and then

- Click **Analyze**, select **Descriptive Statistics,** and then click **Frequencies**. A window will open titled Frequencies.
- Click the **Help** box at the bottom of the window. Another window will open. You will see that Frequencies is highlighted in the left panel, and in the right panel you will see a description of the purpose of the Frequencies procedure and an example.

Here's another example.

- Click **Analyze** on the Main menu, and click **Missing Value Analysis**. A window titled Missing Value Analysis will open.
- Click the **Help** box at the bottom of the window. Another window will open. You will see Missing Value Analysis highlighted in the left panel, and in the right panel you will see information describing the topic of missing values and an example.

Help is available for every option included in the Analyze menu.

7.7 SUMMARY Δ

In this chapter, you learned how to locate topics using the Topics selection on the Help menu. You worked through a tutorial available when Tutorial is selected on the Help menu, and you also investigated a case study by selecting Case Studies. In addition, you learned that help is available for the options listed when you click **Analyze** on the main menu. These methods of obtaining help will prove to be invaluable as you work through the remaining chapters in this book.

Chapter 8

Creating Basic Graphs and Charts

△ 8.1 Introduction and Objectives

SPSS uses the terms *chart* and *graph* interchangeably, and we will do the same. SPSS offers three methods (procedures) to create graphs from a data set: Chart Builder, Graphboard Template Chooser, and Legacy Dialogs. You can use any one of these methods to create the same graph, but the sequence of steps to create that graph using each method is different.

The term *Legacy* suggests that this method was the original method for creating graphs, which it was. Then came Graphboard, and finally, Chart Builder. We prefer Chart Builder because it is powerful and yet easy to master and use to produce quite stunning graphs. However, many users of SPSS were brought up using Legacy Dialogs, so we will also describe and provide directions for using this method.

We use the same data set to create the same graphs using each method so you can compare the two methods. You should experiment with each method and decide which is best for your purpose. Time and patience are required to become proficient in using either of these methods to produce graphs that adequately depict the information and data you have collected and analyzed.

We will leave it as an exercise for you to investigate the use of Graphboard Template Chooser if you wish to do so.

As your databases increase in size and become more complex, and as your graphs become more sophisticated, you may change your preference for a graphing method. In this chapter, you will create basic graphs. In the next chapter, we will show you how to embellish these graphs to make them worthy of being embedded in your documents. We will explain how to extract your graph and other tables that may contain statistical data and place these in applications such as word processors and spreadsheets.

You will be using two sources of data to create graphs: (1) the database you saved in Chapter 5 titled **class_survey2** and (2) one of the sample files included when SPSS was installed on your computer. Refer to the steps listed in Chapter 3 for directions in locating these sample files.

OBJECTIVES

After completing this chapter, you will be able to

- Describe the steps required to create a graph using Legacy Dialogs
- Describe the steps required to create a graph using Chart Builder
- Use Legacy Dialogs to create various types of graphs
- Use Chart Builder to create various types of graphs
- Compare the Legacy Dialogs process with the Chart Builder process

You have heard the adage that a picture is worth a thousand words. Well, a graph depicting a systematic summary and analysis of data is also worth a thousand words.

8.2 USING LEGACY DIALOGS TO CREATE A HISTOGRAM

A histogram, which is a series of contiguous vertical rectangles, is appropriate for displaying information about variables and a range of scores. Histograms represent continuous data, such as scores on a test. Consequently, histograms are useful for displaying frequency information, with scores on the *x*-axis and frequencies (number of each score) on the *y*-axis. Histograms are especially important because they show the distribution of data.

Let's generate a simple histogram using students' average scores on the two exams included in the file that you saved in Chapter 5 under the name **class_survey2**.

- Click **File**, select **Open**, and click **Data**.
- Locate and open the file titled **class_survey2.sav**.
- Click **Graphs**, select **Legacy Dialogs**, and then click **Histograms**. A window will open titled Histogram.
- Click **Average Points On Exams** in the left panel, and click the top right arrow to place it in the Variable box.
- Click **Display normal curve**, then click **OK**.

Your graph will be displayed in the Output Viewer as shown in Figure 8.1.

Figure 8.1 Histogram: Students' Exam Scores

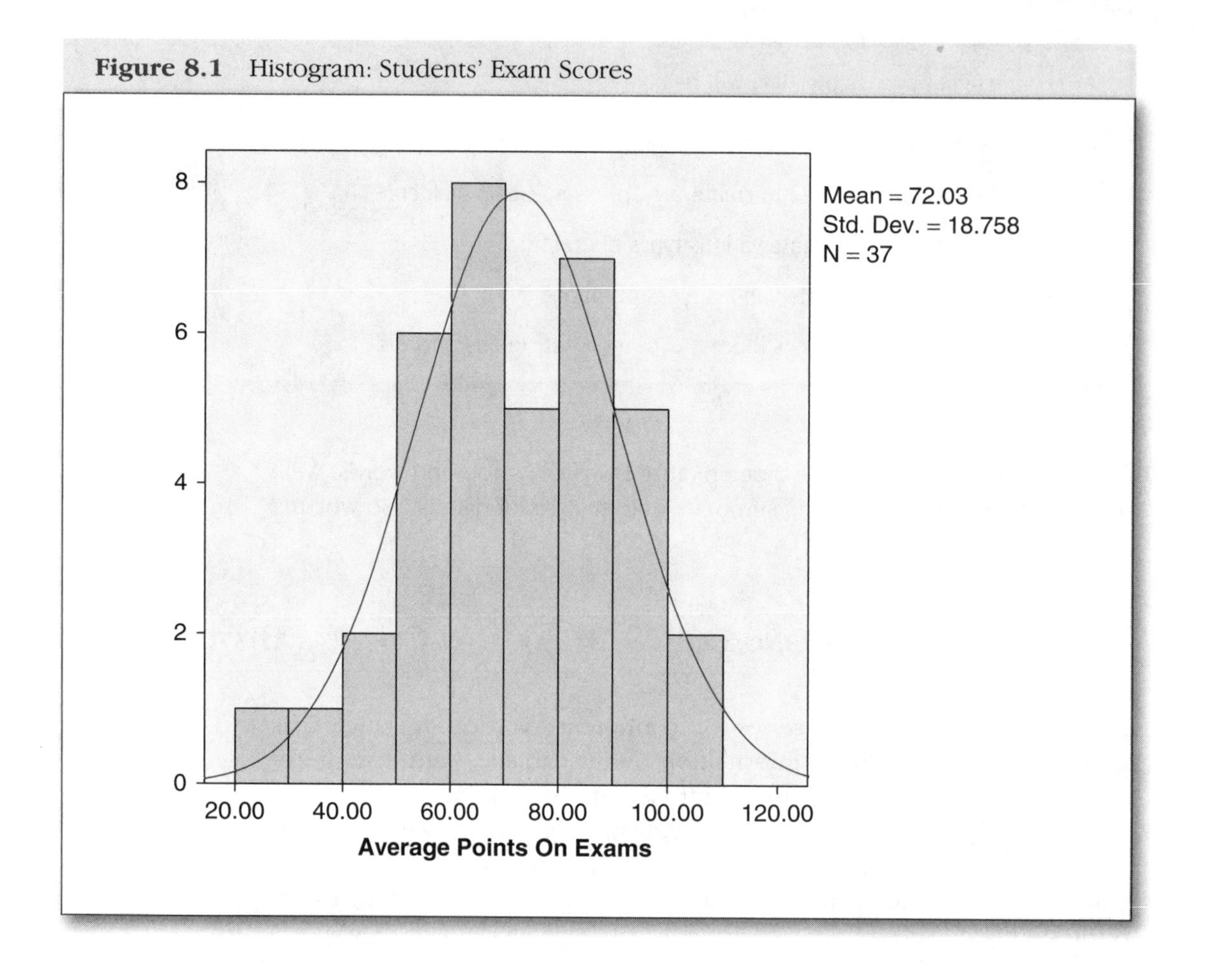

Scores are shown on the *x*-axis and frequencies (the number of each score) are shown on the *y*-axis. You can see by inspecting the graph that the greatest frequency of scores is in the 60 to 70 range. What range has the lowest frequency of scores?

You will see a bell-shaped graph (the normal curve) superimposed over the histogram indicating how well the distribution of scores approximates the normal curve. You will learn about the normal curve in later chapters.

8.3 USING CHART BUILDER TO CREATE A HISTOGRAM △

Let's create the same histogram using Chart Builder.

- Click **Graphs**, and select **Chart Builder**. A window titled Chart Builder will open. At the bottom of the window make certain **Gallery** is selected.
- Click **Reset** (if you make a mistake following the steps to build the graph, simply click **Reset** again and start over).
- Click **Histogram** in the left panel.
- Click the first graph (it looks like a histogram) in the right panel and drag it up to the large white panel at the top-right portion of the window. You will see a simulated graph appear in this panel with boxes representing the *x*-axis and the *y*-axis. Click **Average Points On Exams** in the Variables panel and drag it to the *x*-axis box.
- In the right panel titled Element Properties, click **Display normal curve** and then click **Apply**.
- Click **OK**.

The same graph you created using Legacy Dialogs will be displayed in the Output Viewer. If you make a mistake along the way, simply click **Reset** and start over.

We will describe the use of the other options listed in the Element Properties panel in Chapter 9.

8.4 USING LEGACY DIALOGS TO CREATE A BAR GRAPH △

A bar graph is similar to a histogram, where the heights of bars (rectangles) represent frequencies. But the horizontal scale is NOT continuous. Bar graphs are often used to display frequency data for group variables. Bar graphs have the same appearance as histograms, but the bars do not touch. The *x*-axis represents separate groups and not a continuous range of scores.

Let's create a bar graph depicting the results of the Morning and Afternoon Class on the average scores.

- Click **Graphs**, select **Legacy Dialogs**, and then click **Bar**. A window titled Bar Charts will open.
- Click **Simple** and then click **Summaries for groups of cases**.
- Click **Define**. A window titled Define Simple Bar Summaries for Groups of Cases will open.
- Click **Morning or Afternoon Class** and click the right arrow to place it in the Category Axis box.
- Click **Other Statistic** (e.g., mean).
- Click **Average Points On Exams** and then click the right arrow to place it in the Variable box.
- Click **OK**.

The graph is shown in Figure 8.2.

Figure 8.2 Bar Graph: Morning or Afternoon Class and Students' Exam Scores

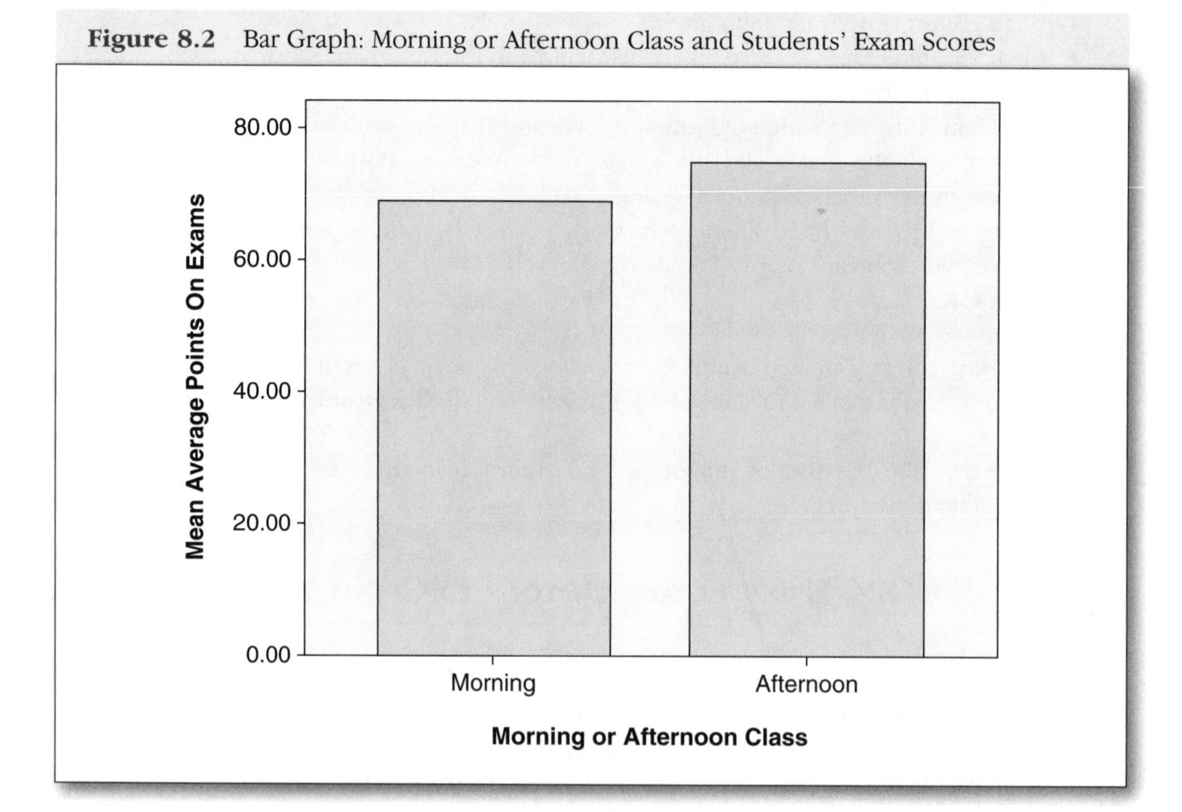

Notice that the *y*-axis does NOT indicate frequencies. Instead, it represents the average (mean) score for each group, Morning or Afternoon Class. That is, the length of each bar represents the average score for each group. Consequently, a bar graph of this type is often called "A Bar Graph of Means."

Inspect the graph. Did the morning class outscore the afternoon class?

8.5 USING CHART BUILDER TO CREATE A BAR GRAPH △

Let's create the same graph using Chart Builder.

- Click **Graphs** and then click **Chart Builder**. When the Chart Builder window opens make certain **Gallery** is selected.
- Click **Reset** (if you make a mistake following the steps to build the graph, simply click **Reset** again and start over).
- Click **Bar** and drag the first **graph** icon up to the large panel on the top right of the window.
- Click **Morning or Afternoon** in the Variables panel and drag it to the *x*-axis.
- Click **Average Points On Exams** and drag it to the *y*-axis (count).
- Click **OK**.

Inspect the graph, which should be exactly the same as the graph you created using Legacy Dialogs.

Let's create one more bar graph depicting how the females did compared with the males.

- Click **Graphs** and then click **Chart Builder**. Make certain **Gallery** is selected.
- Click **Reset** (if you make a mistake following the steps to build the graph, simply click **Reset** again and start over).
- Click **Bar** and drag the second graph icon up to the large panel on the top right of the window.
- Click **Morning or Afternoon Class** in the Variables panel and drag it to the *x*-axis.
- Click **Average Points On Exams** and drag it to the *y*-axis (count).

- Click **Gender** and drag it to the box in the upper right corner that reads "Cluster on x."
- Click **OK**.

The graph is shown in Figure 8.3.

Figure 8.3 Bar Graph: Morning or Afternoon Class, Students' Exam Scores, and Gender

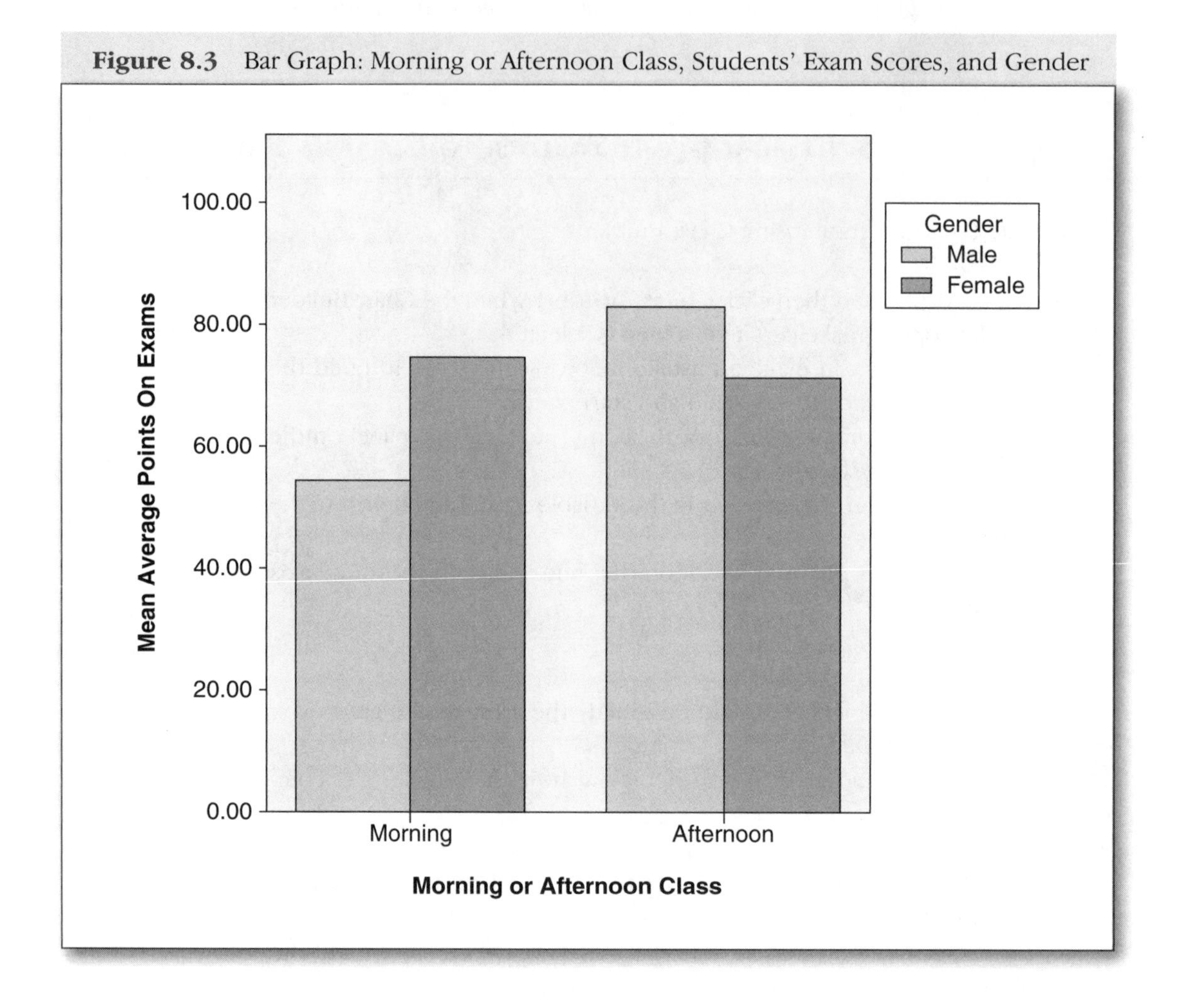

Inspect the graph. Did females outperform males in both classes?

△ 8.6 USING LEGACY DIALOGS TO CREATE A LINE GRAPH

A line graph, probably the most common type of graph, is appropriate for displaying a visual summary of categorical values. Line graphs are easy to read and interpret and are useful in showing trends. Let's investigate a student's predicted grade after taking the first exam. Let's use Legacy Dialogs to create a line graph using "students' predicted grade" on the x-axis and "average points on exams" on the y-axis.

- Click **Graphs** and select **Legacy Dialogs**, and then click **Line**.
- Click **Simple** and then click **Summaries for groups of cases**.
- Click **Define**. A window titled Define Simple Bar Summaries for Groups of Cases will open.
- Click **Students Predicted Final Grade** and click the **right arrow** to place it in the Category Axis box.
- Click **Other Statistic** (e.g., mean).
- Click **Average Points On Exams** and then click the right arrow to place it in the Variable box.
- Click **OK**.

The graph is shown in Figure 8.4.

Figure 8.4 Line Graph: Predicted Grade and Students' Exam Scores

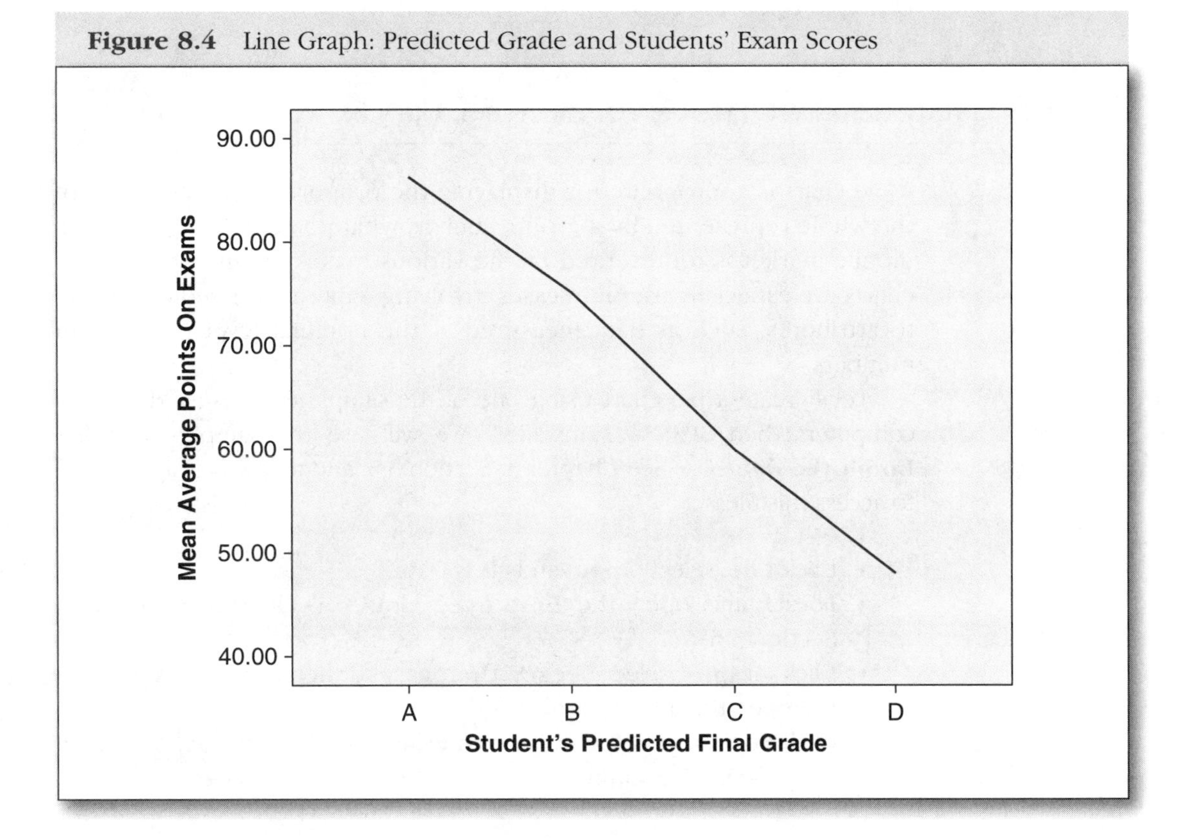

Inspect the graph. Did the students' predicted grades based on their scores agree with your expectations?

△ 8.7 USING CHART BUILDER TO CREATE A LINE GRAPH

Let's create the same line graph using Chart Builder.

- Click **Graphs** and then click **Chart Builder**. Make certain **Gallery** is selected.
- Click **Reset** (if you make a mistake following the steps to build the graph, simply click **Reset** again and start over).
- Click **Line**, and drag the first graph icon in the lower panel up to the upper right panel. Drag **Students Predicted Final Grade** to the *x*-axis and drag **Average Points On Exams** to the *y*-axis (count).
- Click **OK**.

The graph should look exactly like the graph you created using Legacy Dialogs. Inspect the graph.

△ 8.8 USING LEGACY DIALOGS TO CREATE A PIE CHART

A pie chart is appropriate for displaying the proportion (percentage) of the whole represented by a group, such as what percentage of the automobile market is represented by the various makes of automobiles. Pie charts are especially useful in cases involving data categorized according to attributes, such as data measured at the nominal level rather than numbers.

Let's create a pie chart using one of the sample files installed in your computer when SPSS was installed. We will use the sample file titled **Employee data.sav**. See Chapter 3, Section 3.5, and follow the steps listed to access this file.

- Click **File**, select **Open**, and click **Data**.
- Locate and open the **Employee data.sav** file (see Chapter 3, Section 3.5).
- Click **Graphs**, select **Legacy Dialogs**, and then click **Pie**. A window will open titled Pie Charts.
- Click **Summaries for groups of cases** and click **Define**. A window titled Define Pie summaries for Groups of Cases will open.

- Click **Educational Level** and then click the **right arrow** to place it in the Define Slices by box.
- Click **OK**.

The chart is shown in Figure 8.5. The colored slices depict the level of education of the employees in years.

Figure 8.5 Pie Chart: Employees' Level of Education in Years

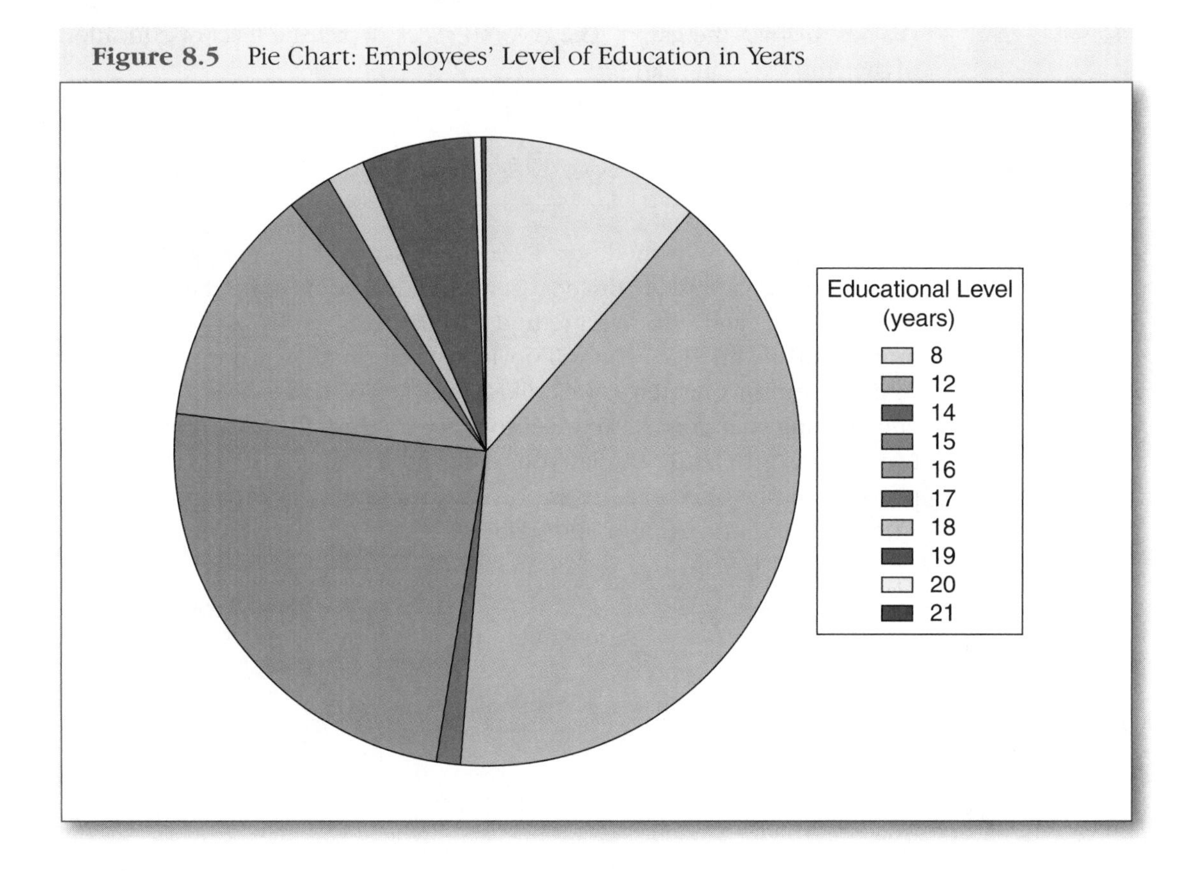

8.9 USING CHART BUILDER TO CREATE A PIE CHART △

Let's use Chart Builder to create the same pie chart.

- Click **File**, select **Open**, and click **Data**.
- Locate and open the **Employee data.sav** file.
- Click **Graphs** and then click **Chart Builder**. Make certain **Gallery** is selected.

- Click **Reset** (if you make a mistake following the steps to build the graph, simply click **Reset** again and start over).
- Click **Pie/polar**. Drag the pie graph icon from the lower panel up to the large panel. Drag **Educational Level** to the *x*-axis.
- Click **OK**.

The graph should look exactly like the graph you created using Legacy Dialogs. Inspect the graph. The colored slices depict the level of education of the employees in years.

△ 8.10 SUMMARY

In this chapter, you learned how to create basic graphs using two methods: Legacy Dialogs and Chart Builder. Chart Builder seems to be the easier and more efficient and effective method for creating graphs. As you become more familiar with SPSS, you will obviously determine which method works best for you. In Chapter 9, we will, as usual, give hands-on directions for embellishing your graphs. We will also show you how to create more sophisticated graphs to better display your data and their purpose. Finally, we will provide specific directions for exporting your graphs to other applications such as a word processor or spreadsheet.

Chapter 9

Editing and Embellishing Graphs

9.1 Introduction and Objectives △

In Chapter 8, you used Legacy Dialogs and Chart Builder to create some basic graphs of various types. In this chapter, you will use Chart Builder to create more sophisticated graphs and Chart Editor to edit these graphs and others you may have already created in your work with SPSS. Legacy Dialogs may also be used to create and edit such graphs, but the process is not as simple and intuitive as using Chart Builder.

You may feel that, after creating a graph, there is no recourse for making changes. Chart Editor will come to your rescue and help you make changes to your graphs such as adding titles, borders, color, shading, annotations, footnotes, and much more. Along the way, you can have some fun messing with different options just to see how they look in the final rendition of your graphs.

After creating and editing your graphs, we will describe the steps necessary for you to extract these from the SPSS Output Viewer and place them in word processing documents, spreadsheets, and other documents to give them a professional look.

You will be using two sources of data to create and edit graphs: (1) the database you saved in Chapter 5 titled **class_survey2** and (2) one of the sample files included when SPSS was installed on your computer. Refer to the steps listed in Chapter 3 for directions in locating these sample files.

Note: The various colors are not displayed in the graphs in this chapter, but they will be displayed on your computer screen.

OBJECTIVES

After completing this chapter, you will be able to

Create a basic graph using Chart Builder

Describe the features available on the Chart Editor

Edit a simple graph using the Chart Editor

Edit a three-dimensional graph using the Chart Editor

Export a graph to your document

△ 9.2 CREATING A BASIC GRAPH

You did this in Chapter 8, so this exercise will serve as a review and refresher regarding creating charts. You will use the **class_survey2** database.

- Click **File**, select **Open**, and click **Data**. Locate and open the file titled **class_survey2.sav**.
- Click **Graphs** and then click **Chart Builder**. Make certain **Gallery** is selected.
- Click **Reset** (if you make a mistake following the steps to build the graph, simply click **Reset** again and start over).
- Click **Line** and drag the first graph icon in the bottom panel to the large panel at the top of the window.
- Click **Students Predicted Final Grade** in the Variable box and drag it to the *x*-axis.
- Click **Average Points On Exams** and drag it to the *y*-axis (count).
- Click **OK**.

The graph is shown in Figure 9.1.

This is one of the graphs you created in Chapter 8. Now you will have an opportunity to embellish it using the Chart Editor as shown in Figure 9.2. Your graph appears in the Output Viewer. Double click on the graph, and the Chart Editor opens as shown in Figure 9.2.

The Chart Editor has a Main menu as shown at the top of the window. Click on each item on the menu to see what options are available. For example if you click **View**, you will see the following options: Status Bar, Edit Toolbar, Options Toolbar, Element Toolbar, Format Toolbar, and Large Balloons.

Figure 9.1 Basic Line Graph: Grade Prediction Based on Exams

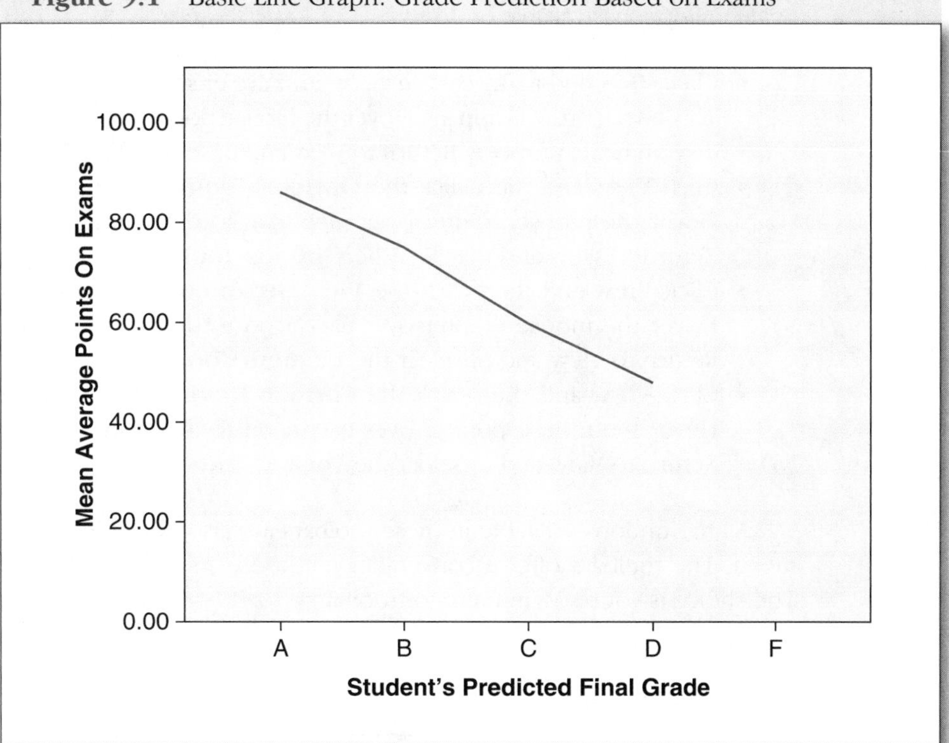

Figure 9.2 Chart Editor

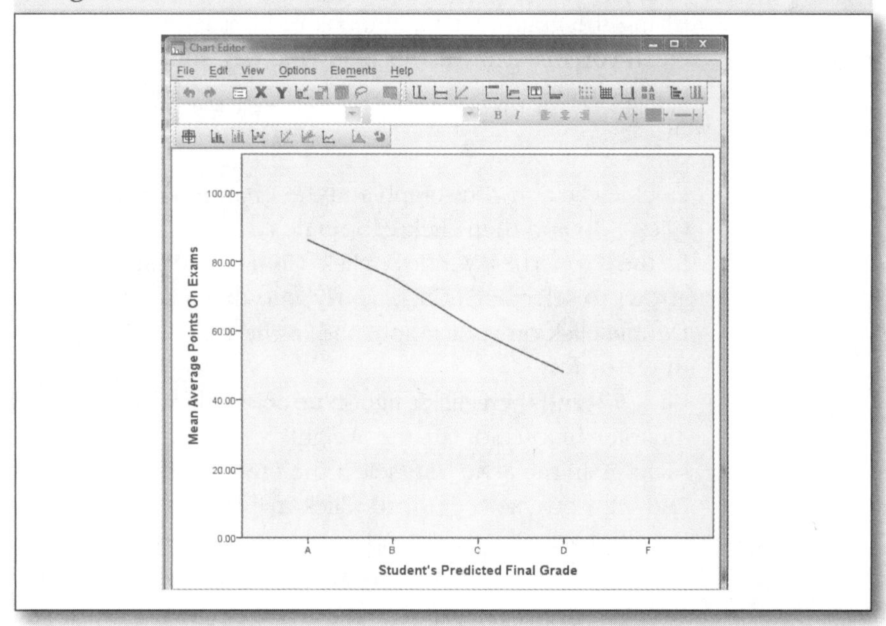

After you complete your tour of the options available on the Main menu, investigate the options available on **View**.

- Click **View** and make certain all options are unselected. Click the **Edit Toolbar.** Icons will appear. Hover the mouse pointer over each icon to determine its purpose. Return to View and unselect the **Edit Toolbar.**
- Click **View** and then click the **Options Toolbar**. Icons will appear. Hover the mouse pointer over each icon to determine its purpose. Return to View and unselect the **Options Toolbar**.
- Click **View** and then click the **Element Toolbar**. Icons will appear. Hover the mouse pointer over each icon to determine its purpose. Return to View and unselect the **Element Toolbar**.
- Click **View** and then click the **Format Toolbar**. Icons will appear. Hover the mouse pointer over each icon to determine its purpose. Return to View and unselect the **Format Toolbar**.

All the options available in these toolbars are also available on the Main menu. The toolbars offer a convenient method of rapidly editing a graph. The choice is yours: Main menu or toolbars!

△ 9.3 EDITING A BASIC GRAPH

First let's use the Chart Editor to change the size of your graph to 4 inches. One inch equals 72 points, so you will resize to approximately 288 points. After resizing the graph, let's add a colored border, a colored background, and a title. If you make any errors along the way, simply click **Edit** and then click **Undo** as many times as you wish until you get back to the point where all is well.

- Double click on your graph and the Chart Editor will open.
- Click **Edit** and then click **Properties**.
- In the Properties Window, click **Chart Size** and use the up-down arrows to select 285. Click **Apply**, and then click **Close**.
- Double click on your graph, and in the Properties Window click **Fill and Border**.
- Click **Fill** and then click a light-blue color. Click **Border** and then click a darker blue color. In the Weight box, use the up-down arrow to select **3**. In the Style box, select the **straight horizontal line**. In the End Caps box, select **round**. Click **Apply**.
- Double click the *y*-axis and in the Properties Window select **Text Style**. Use the up-down arrows to change the Preferred Size to 10.

Scroll in the Family box and select **Times New Roman**. In the Style box, use the up-down arrows to select **Bold**. Click **Apply**.

- Double click the *x*-axis and follow the same directions given for the *y*-axis. Click **Close**.
- Click **Options** on the Chart Editor menu and then click **Title**, and a rectangle will open above the graph in which you can type a title for your graph. Type "Predicted Score Based On Exam Averages." Choose **Times New Roman**, size **14**, and **Bold**. Click **Apply** and then click **Close**.
- Click **File** and then click **Close** on the Chart Editor menu, and your graph will be placed in the Output Viewer.

Your graph should look something like the one in Figure 9.3.

Figure 9.3 Edited Line Graph: Grade Prediction Based on Exams

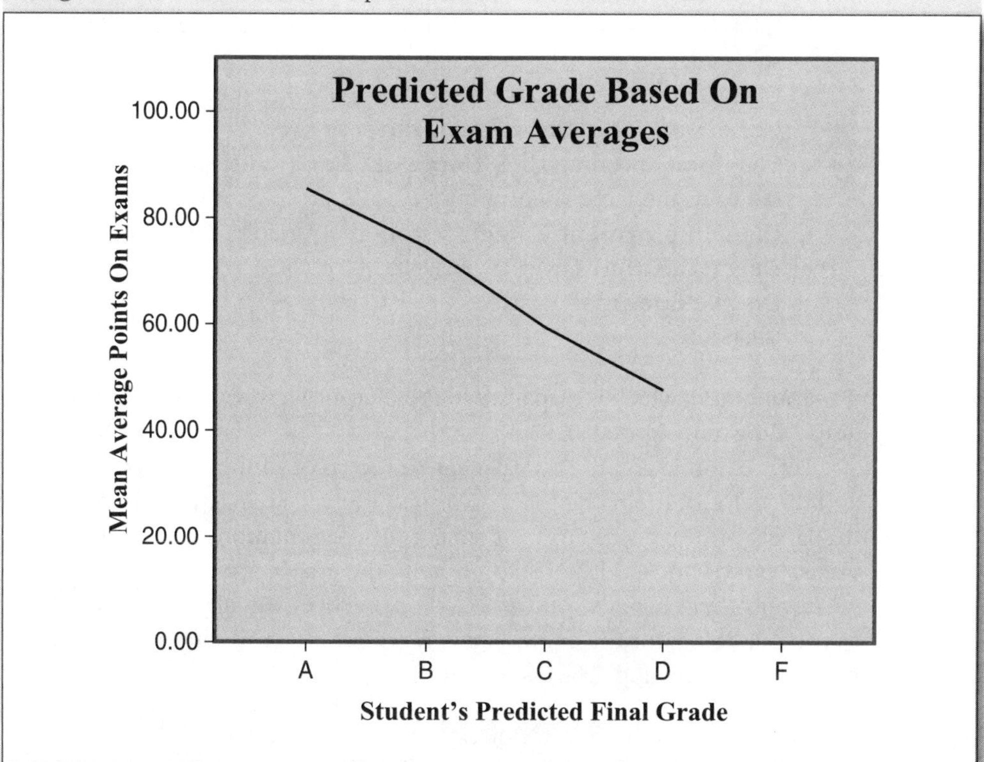

It is best to keep graphs simple and uncomplicated. Chart Editor offers many features, and the novice is prone to overdo it to the point where a graph becomes so cluttered with features and details as to be uninterpretable.

Place your original graph back in the Chart Editor, play with all the various settings and windows, and come up with your own creation.

Note: We discovered that the "Title" and other options such as "Footnote" do not always work properly if applied after one has made other changes to a graph. However, if the "Title" and "Footnote" options are requested before any other changes have been made to a graph, they seem to work appropriately (also see Section 9.4).

△ 9.4 EDITING A THREE-DIMENSIONAL GRAPH

In the following exercise, you will create a three-dimensional graph using one of the sample files. Then you will edit this graph in the Chart Editor.

- On the Main menu, click **File**, select **Open**, and click **Data**.
- Locate and open the sample file titled **Employee data**.
- Click **Graphs** and then click **Chart Builder**. Make certain **Gallery** is selected.
- Click **Reset** (if you make a mistake following the steps to build the graph, simply click **Rese**t again and start over).
- Click **Bar** and drag the **Simple3-D Ba**r (fourth icon in the lower pane) to the large window above.
- Drag **Employment Category** from the Variables pane to the *x*-axis.
- Drag **Education Level** to the *y*-axis (count).
- Drag **Gender** to the *z*-axis.
- Click **OK**.

Your graph, which is displayed in the Output Viewer, should look like the graph shown in Figure 9.4.

Let's edit this graph. You will change the color of the two vertical panels and the color of the rectangular columns. You will change the orientation of the labels on the *z*-axis. You will change the orientation of the title on the *y*-axis. You will add a title and a footnote to the graph. And then you can make whatever other changes you wish just for practice. Actually, you do not need to use our colors for the graph. Use any colors you wish, but try to make the colors complementary. Have some fun!

- In the Output Viewer, double click on the **graph**.
- Double click the left vertical panel. Notice that a border surrounds the part of the graph you have selected. If the border is not surrounding the left panel, click outside the graph and then click the left panel again. In the Properties Window, click **Fill and Border**. Click **Fill** and click the red box color. Click **Border** and then click the black box. Click the **Weight** box and use the up-down arrows to select **3**. Click

the **Style** box, and select the straight horizontal bar. Click the **End Caps** box and select round. Click **Apply**.

- Double click the right vertical panel. In the Fill and Border window, click **Fill** and click the same red box color. Click **Edit**, and a screen titled Choose a Color will open. Click on a red color significantly lighter than on the first panel. Click **OK**.
- Click **Border**, and follow the directions given for the left panel. Click **Apply**.
- Double click the base of the graph and follow the same directions as given for the left panel for color and border. Click **Apply**.
- Double click the rectangular columns. In the Properties Window, select **Fill and Border.** Click **Fill** and click the yellow box color. Click **Border** and follow the directions given for the other panels, except select **1** for weight. Click **Apply**.
- Double click **Educational Level (Years)** on the *y*-axis and in the Properties window click **Text Layout**. In the Orientation box, click **Horizontal**. Click **Apply**.

Figure 9.4 Three-Dimensional Graph Summarizing Employment

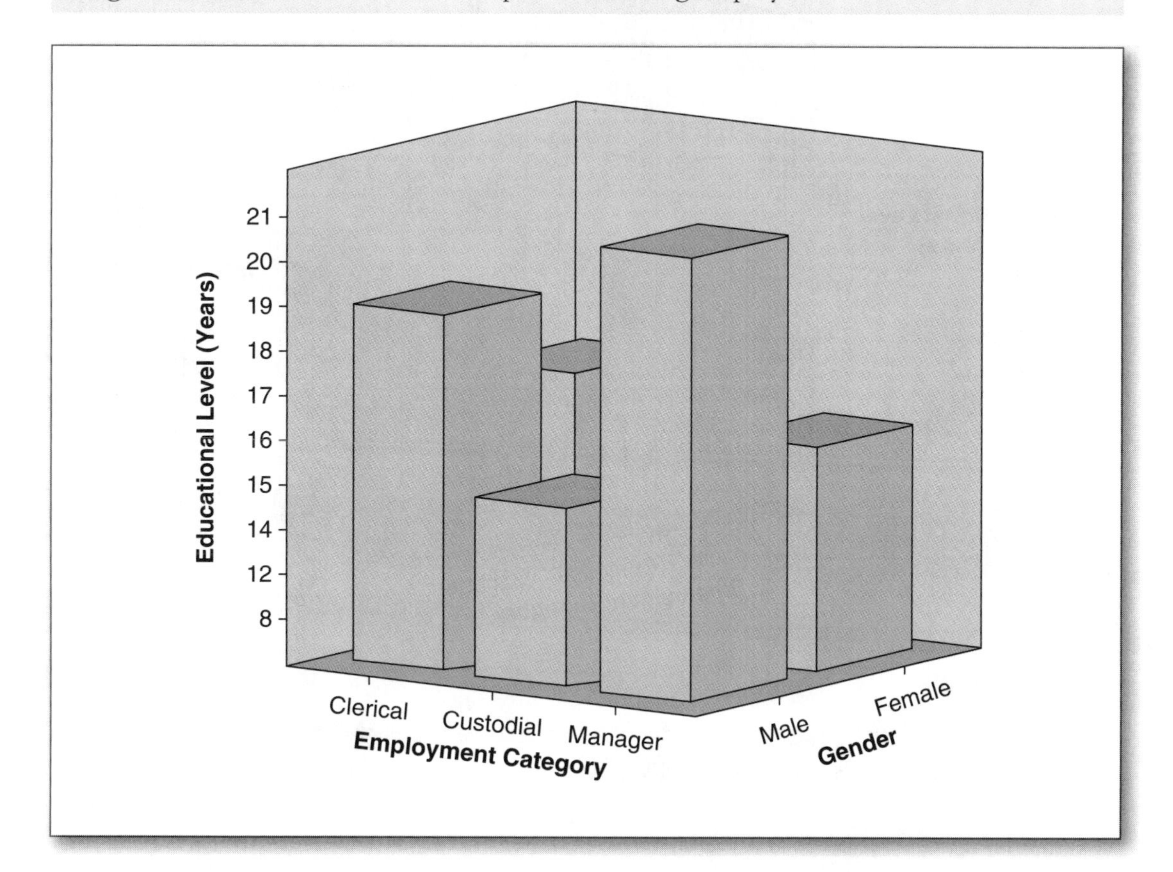

- Click **Male Female** on the z-axis. In the Properties Window, click **Labels and Ticks**. Click **Display Labels** and select **Diagonal**. Click **Apply**.
- Click **Options** on the Chart Editor menu and then click **Title**. A rectangle will open above the graph in which you can type a title for the graph. Type **Employment Summary**. Press the Enter key on the keyboard.
- Click **Options** and then click **Footnote.** Type "Data for this graph were provided by a large company."
- In the Chart Editor, click **File** and then click **Close**. Your graph will be placed in the Output Viewer.

Your graph should look like that shown in Figure 9.5 unless you became more creative and really embellished the graph!

Figure 9.5 Edited Three-Dimensional Graph Summarizing Employment

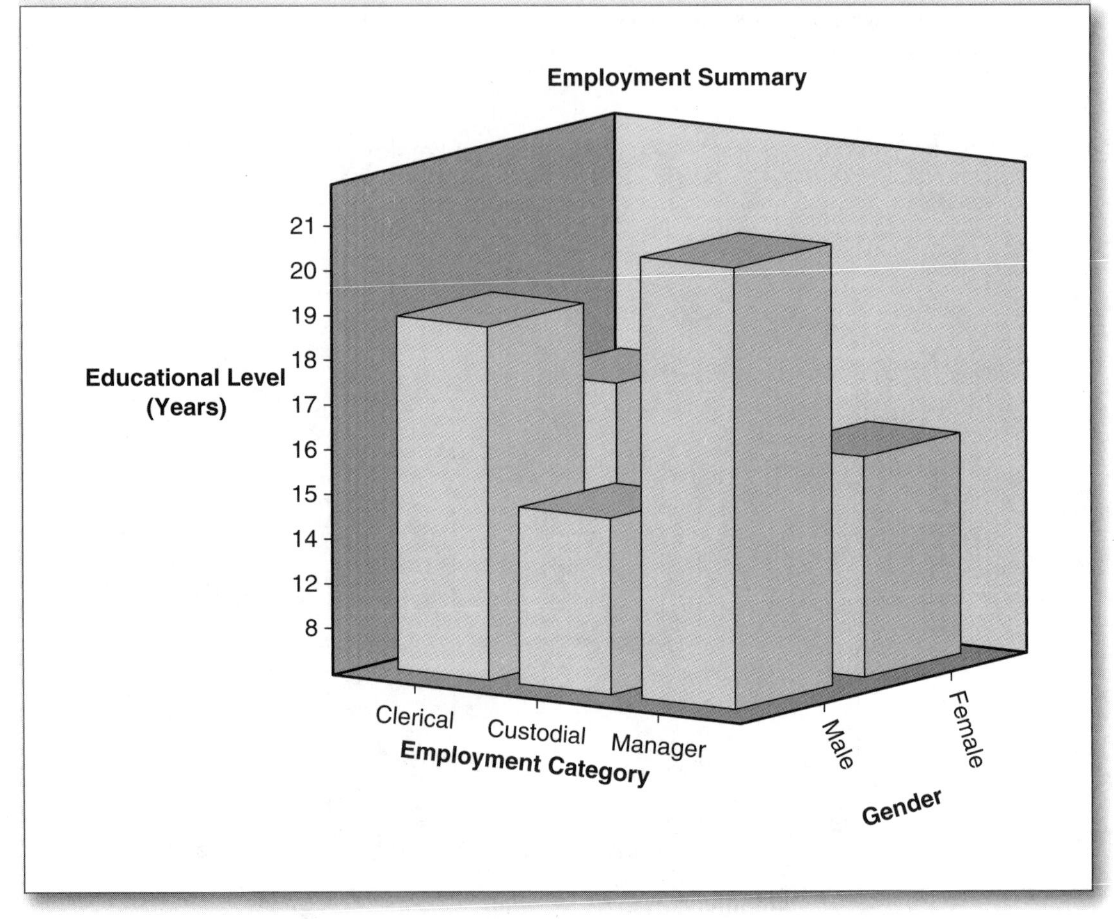

Note: Data for this graph were provided by a large company.

9.5 EXPORTING GRAPHS TO DOCUMENTS △

Exporting graphs to your documents is a simple process. With the graph in the Output Viewer, click the graph, and on the Main menu of the Output Viewer, click **Edit** and then click **Copy**. The file will be copied to the computer's clipboard. Open your document, such as a word processing document or a spreadsheet, select the position in the document where you wish to place the graph, and then paste the graph in the document. Once the graph is in your document, you may resize it to fit if necessary.

9.6 SUMMARY △

In this chapter, you learned how to edit graphs to ensure that they are not only highly representative of the data but also appealing and engaging to the viewer. In Chapter 10, you will learn techniques and procedures for printing the results of your analyses using SPSS, including printing of data, information regarding variables, and tables.

Chapter 10

Printing Data View, Variable View, and Output Viewer Screens

△ 10.1 Introduction and Objectives

Printing raw data and the information that describes your variables (input) and the results of data analysis (output) is a major task made easy (most of the time) by SPSS. This chapter continues the theme and purpose of the prior chapters, namely, how can we make our observations (the data) more understandable and therefore useful? You will expand on this theme by printing the results of your analysis, which can then be used in statistical reports or perhaps for a more careful and relaxed examination while having a cup of coffee with a friend. There are three major sections in this chapter: (1) How do you print data from the Data View screen? Printing a copy of the data you entered can be very helpful when checking the accuracy of your work. You are also shown how to select and print a portion of your data. (2) How do you print variable information? Having a printed version of the information used to describe your variables can be very useful as you proceed with the task of data analysis. A printed summary of variable information also serves as a "codebook" by providing a list of all variables and their codes for your categorical variables. Another purpose is that the codebook makes it easier to share your work with a colleague for comment and/or discussion. (3) How do you print tables, graphs, and other analysis from the Output Viewer screen? These are the critical portions of your analysis that are intended to provide a succinct summary of all the work. Printing and embellishing selected portions of your output provides a convenient communication method with colleagues and the world.

OBJECTIVES

After completing this chapter, you will be able to

- Print all entered data as displayed in the Data View screen
- Print descriptive information for variables from the Variable View screen
- Print output such as tables and graphs from the Output Viewer screen
- Use the mouse to select and print data

10.2 PRINTING DATA FROM THE VARIABLE VIEW SCREEN △

Printing data that have been entered into the Data View screen in a format easy to interpret is not always a simple task. In general, when your database has more than six variables and more than 36 cases, SPSS spreads the printing over multiple pages. The printing is easy, one click, but the collation and interpretation of a data-printing job can be very challenging. Let's use **class_survey1.sav** to illustrate what we mean when we say that the collation and interpretation are difficult. The class survey database contains seven variables and 37 cases.

- Start SPSS and click **Cancel** in the SPSS Statistics opening window.
- Click **File**, select **Open**, and click **Data**.
- In the file list, locate and click **class_survey1.sav** (database opens).
- Click the **Data View** tab if it has not already been clicked.
- Click **File**, and then click **Print Preview** (the Data Editor screen opens; see Figure 10.1).

Figure 10.1 Print Preview for the Data View Screen: First of Four Pages

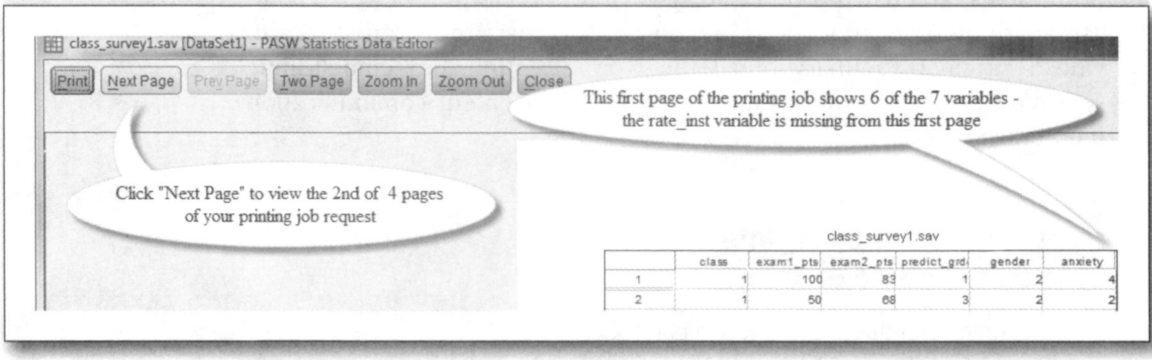

- Click the **Next Page** (shown in Figure 10.1), which opens another page that displays the seventh variable (rate_inst), creating another entire page as shown in Figure 10.2.

Figure 10.2 Print Preview for the Data View Screen: Second of Four Pages

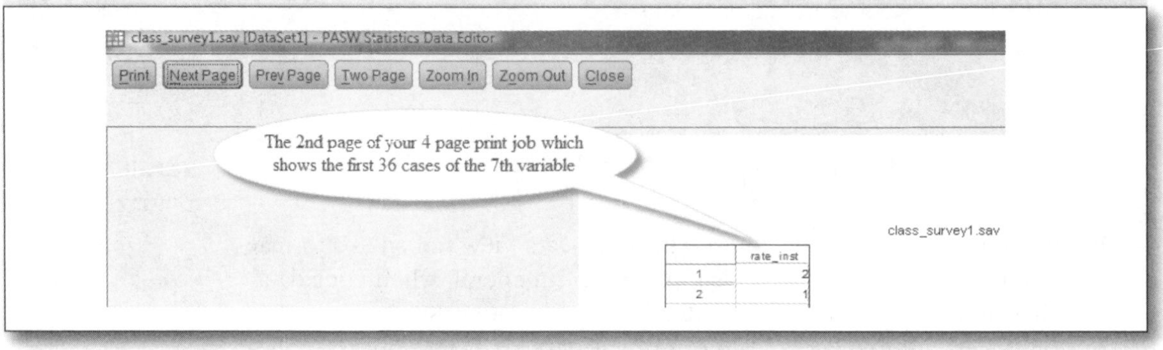

- Clicking **Next Page** brings up the third page—the 37th case, as shown in Figure 10.3.

Figure 10.3 Print Preview for the Data View Screen: Third of Four Pages

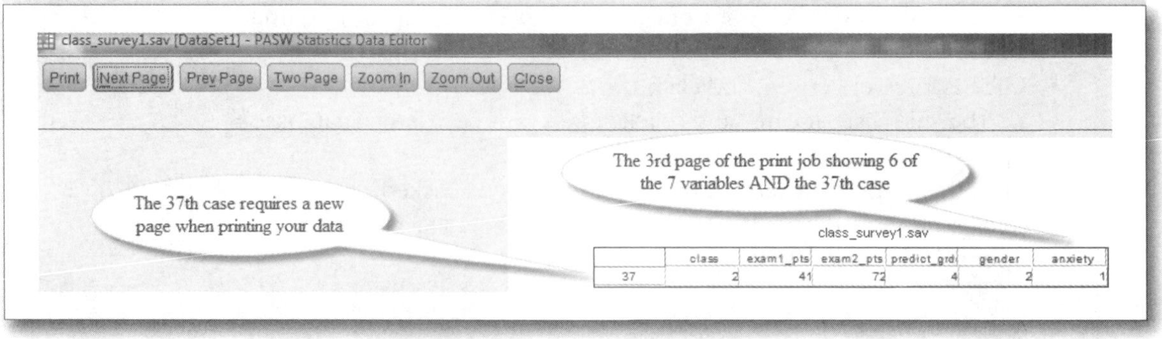

- Clicking **Next Page** again brings up the final page of your printing job (see Figure 10.4).

Figure 10.4 Print Preview for the Data View Screen: Fourth of Four Pages

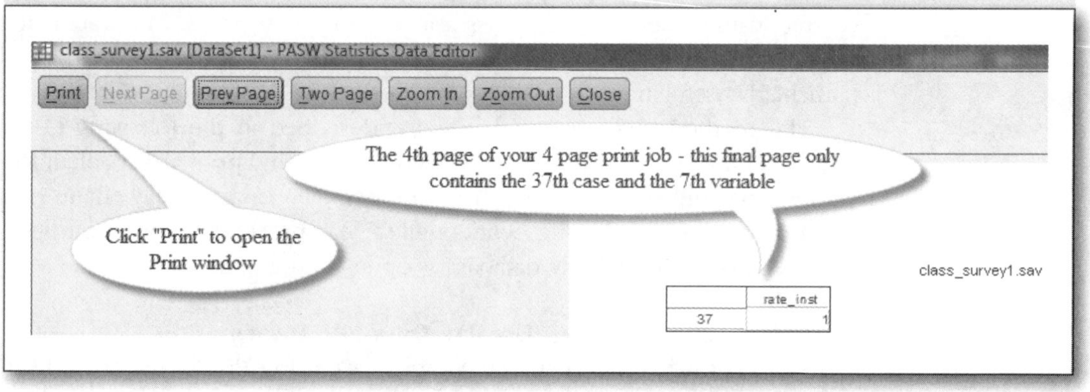

After following this procedure, one can surmise the complexity of a print job having hundreds of cases and dozens of variables. We would like to be able to describe an easy way to assemble the four pages (Figures 10.1–10.4) into one understandable format. Unfortunately, we cannot. If you must have a hard copy of your data, we can only advise you to get out your scissors and cellophane tape. The scissors-and-tape procedure takes time, but keep in mind that you rarely need a hard copy of your data.

To complete the printing job of the four pages as depicted in Figures 10.1 through 10.4, do the following:

- Click **Print** as shown in Figure 10.4, and the Print window opens (see Figure 10.8 later in this chapter).
- Click **OK** (your printer is activated, and your printing job is completed).

Printing a Selected Portion of Your Data

There are occasions when you may wish to print some subset of your database. To print the desired portion, click the mouse and drag the pointer over the desired cases and variables that you wish to print. These areas will automatically be highlighted. Then follow the normal printing sequence.

- Click **File**, and then click **Print**.
- Click **Selection**, and then click **OK**.

The click-and-drag process allows you to save ink and paper if you only require a hard copy of a portion of your data.

△ 10.3 PRINTING VARIABLE INFORMATION FROM THE OUTPUT VIEWER

It is often useful to have a printed copy of the labels, value labels, level of measurement, and other information associated with your variables. There are many ways to print such information. One of these methods (the File method) is particularly useful in that it provides the information in a printable format.

The method recommended by us is described in the following bullet sequence. It provides a quick and easy way to view and print all variable information from any SPSS file. If you have an open file, you can easily obtain variable information by selecting "working file." Let's proceed on the assumption that the **class_survey1.sav** database is open—if not, please open this file.

- Click **File,** and select **Display Data File Information**.
- You now have a side menu choice of **Working File** or **External File**.
- Click **Working File**.
- If the Output Viewer does not automatically open, click **Output** at the bottom of your screen to view Figures 10.5 and 10.6.

The Output Viewer opens with two tables: Variable Information (Figure 10.5) and Variable Values (Figure 10.6). The "Working File" side menu item was selected since this was the file that was currently open and active. You would select External File if you desired information on any other file.

Figure 10.5 Variable Information for the Class Survey Database

Variable Information

Variable	Position	Label	Measurement Level	Role	Column Width	Alignment	Print Format	Write Format
class	1	Morning or Afternoon Class	Nominal	Input	8	Right	F8	F8
exam1_pts	2	Points on Exam One	Scale	Input	8	Right	F8	F8
exam2_pts	3	Points on Exam Two	Scale	Input	8	Right	F8	F8
predict_grde	4	Student's Predicted Grade	Nominal	Input	8	Right	F8	F8
gender	5	Gender	Nominal	Input	8	Right	F8	F8
anxiety	6	Self-rated Anxiety Level	Ordinal	Input	8	Right	F8	F8
rate_inst	7	Instructor Rating	Ordinal	Input	8	Right	F8	F8

Variables in the working file

Figure 10.6 Variable Values for the Class Survey Database

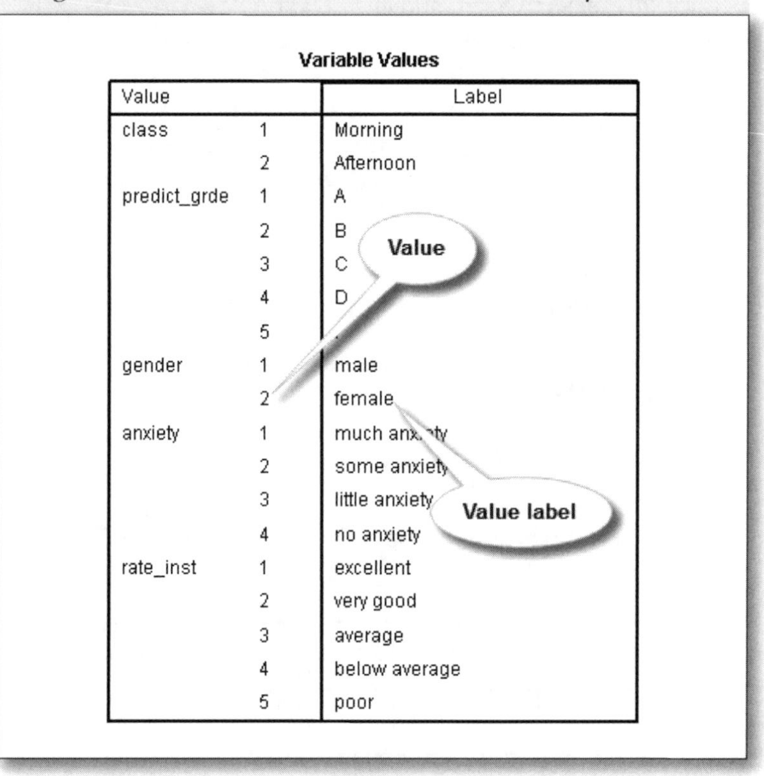

Variable Values

Value		Label
class	1	Morning
	2	Afternoon
predict_grde	1	A
	2	B
	3	C
	4	D
	5	.
gender	1	male
	2	female
anxiety	1	much anxiety
	2	some anxiety
	3	little anxiety
	4	no anxiety
rate_inst	1	excellent
	2	very good
	3	average
	4	below average
	5	poor

To complete the printing operation, do the following:

- Click **File**, and then click **Print**.
- Click **OK**.

The two tables shown in Figures 10.5 and 10.6 provide all the information about the specifications of your variables.

10.4 PRINTING TABLES FROM THE OUTPUT VIEWER △

When printing from the Output Viewer, there are several ways to enhance the appearance of your print job. Of course, if the printing is only for your use, the appearance may not matter. In this case, you may simply wish to print whatever appears in the Output Viewer.

Let's begin this section by generating some output from the **class_survey1 .sav** database.

- Start SPSS, and then click **Cancel** in the SPSS Statistics opening window.
- Click **File**, select **Open**, and click **Data**.
- In the file list, locate and click **class_survey1.sav**.
- Click **Analyze**, select **Descriptive Statistics**, and click **Frequencies**.
- Click **Student's Predicted Grade** and **Instructor Rating** while holding down the computer's **Ctrl** key (this moves variables to the right panel).
- Click arrow, and then click **OK** (Figure 10.7 will appear in the Output Viewer).

Figure 10.7 Output Viewer: Output as It Appears (Rough Draft)

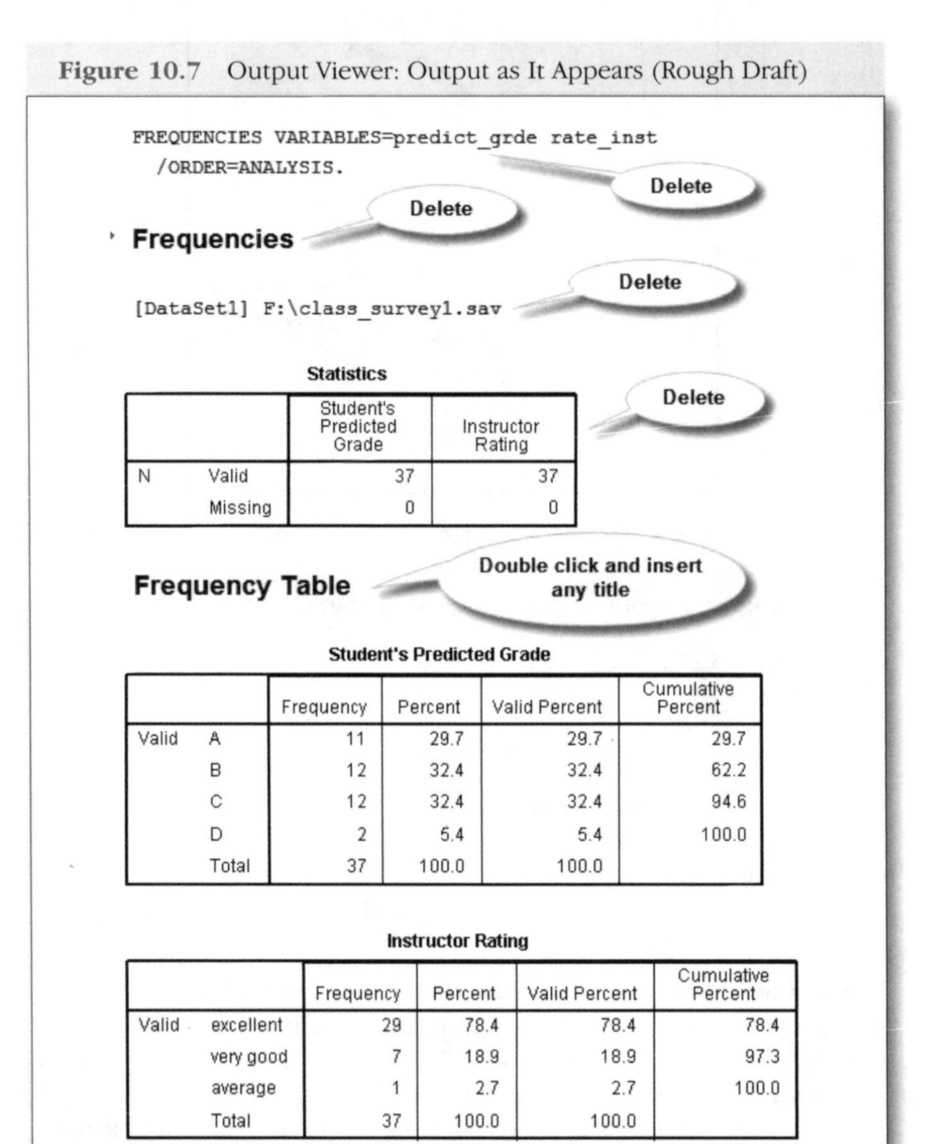

```
FREQUENCIES VARIABLES=predict_grde rate_inst
  /ORDER=ANALYSIS.
```

Frequencies

```
[DataSet1] F:\class_survey1.sav
```

Statistics

		Student's Predicted Grade	Instructor Rating
N	Valid	37	37
	Missing	0	0

Frequency Table

Student's Predicted Grade

		Frequency	Percent	Valid Percent	Cumulative Percent
Valid	A	11	29.7	29.7	29.7
	B	12	32.4	32.4	62.2
	C	12	32.4	32.4	94.6
	D	2	5.4	5.4	100.0
	Total	37	100.0	100.0	

Instructor Rating

		Frequency	Percent	Valid Percent	Cumulative Percent
Valid	excellent	29	78.4	78.4	78.4
	very good	7	18.9	18.9	97.3
	average	1	2.7	2.7	100.0
	Total	37	100.0	100.0	

Return to the Output Viewer, and follow these steps to produce an output more suitable for inclusion in a statistical report. In the Output Viewer, do the following:

- Right click and "delete" all items prior to "Frequency Table."
- Double click **Frequency Table** and type **Statistics Class Survey** and then delete the words **Frequency Table** (Frequency Table is replaced with Statistics Class Survey).
- Click **File**.

Once you click **File**, you will see three printing choices provided on the drop-down menu: (1) Page Setup, (2) Print Preview, and (3) Print. In the majority of cases, you will simply select Print and proceed from there. However, we now briefly discuss the other two choices.

The first option, as mentioned above, is "Page Setup," which provides the opportunity to specify paper size, portrait or landscape orientation, and margin size. The second option, "Print Preview," simply shows what the output will look like once it is printed.

Note: If you have any graphs in the Output Viewer, you will see a fourth printing option, "Page Attributes." Clicking on the **Page Attributes** choice opens the Page Attributes window. You can then add custom headers and footers to your output. By clicking the **Options** tab in this window, you may also specify the number of graphs printed on each page. This printing option can be very useful when printing a large number of graphs. For this exercise, we select the "Print" option.

- Click **Print** (Print window opens as shown in Figure 10.8).

Figure 10.8 Print Window

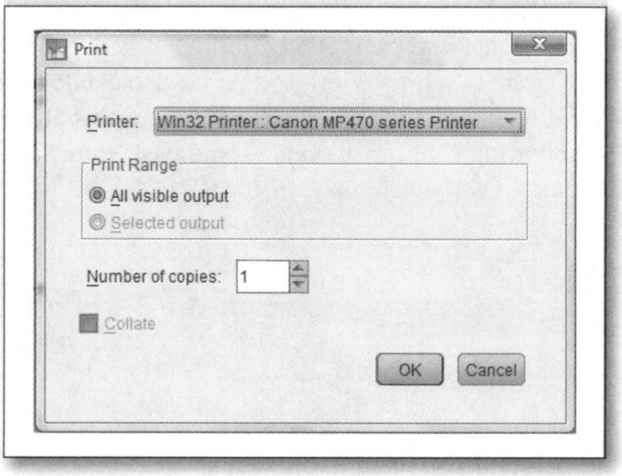

- Click **OK**.

The final, and enhanced, printed output is shown in Figure 10.9, and it is ready to be included in a statistical report.

Figure 10.9 Enhanced Printed Class Survey Output

Statistics Class Survey

Student's Predicted Grade

		Frequency	Percent	Valid Percent	Cumulative Percent
Valid	A	11	29.7	29.7	29.7
	B	12	32.4	32.4	62.2
	C	12	32.4	32.4	94.6
	D	2	5.4	5.4	100.0
	Total	37	100.0	100.0	

Instructor Rating

		Frequency	Percent	Valid Percent	Cumulative Percent
Valid	excellent	29	78.4	78.4	78.4
	very good	7	18.9	18.9	97.3
	average	1	2.7	2.7	100.0
	Total	37	100.0	100.0	

△ 10.5 SUMMARY

In this chapter, you learned basic SPSS printing procedures. You are now able to print data from the Data View screen, variable information from the Variable View screen, and various types of SPSS outputs given in the Output Viewer. The following chapter provides some basic information on descriptive statistics and how they are used by SPSS.

CHAPTER 11

BASIC DESCRIPTIVE STATISTICS

11.1 INTRODUCTION AND OBJECTIVES △

This chapter presents basic statistical tools available in SPSS that are used to make data more understandable. The better we understand our data, the more useful they become in assisting us in discovering patterns and making informed decisions.

We begin with a definition of descriptive statistical analysis. Descriptive statistical analysis is any statistical or mathematical procedure that reduces or summarizes numerical and/or categorical data into a form that is more easily understood.

Descriptive statistics is that branch of statistics where data have been collected and you now wish to describe that data. Contrast this with the other major branch of statistics, known as inferential statistical analysis. Inferential statistics uses sample data to generate approximations of *unknown* values in the population. In other words, it goes beyond descriptive statistical analysis. Descriptive statistical analysis does not mean that inferential analysis is somehow superior but only that it serves a different purpose. In fact, many inferential techniques require that you first conduct a descriptive analysis of the sample data obtained from a population—more on this aspect of descriptive analysis in Section 11.4.

There are many descriptive statistical techniques designed to accomplish data reduction and summarization. For instance, you may find it useful to generate frequency tables, graphs, and scatter plots. Graphs and tables were covered in Chapters 8 and 9; therefore, tables are emphasized in this chapter. When conducting descriptive statistical analysis, you may wish to calculate the mode, median, mean, standard deviation, variance, range, skewness, or kurtosis. The current chapter gives you the skills needed to use the power of SPSS to describe any distribution of data (a bunch of numbers). Another important descriptive statistic is the correlation coefficient, which describes the strength of a relationship between two variables (bivariate analysis). Bivariate analysis is covered in Chapter 19, whereas this chapter covers those methods used to describe single variables, one at a time (univariate analysis).

OBJECTIVES

After completing this chapter, you will be able to

- Describe and interpret measures of central tendency
- Generate and interpret frequency tables
- Describe and interpret measures of dispersion (variability)
- Generate and interpret descriptive statistical tables
- Describe and summarize variables that are measured at nominal, ordinal, and scale levels
- Determine if a variable's values approximate the normal distribution

△ 11.2 MEASURES OF CENTRAL TENDENCY

Measures of central tendency are used when we wish to describe one variable at a time. An example of such data description would be the class survey that we have used in prior chapters. Think in terms of the average score on the first exam, the number of females and males, or perhaps level of anxiety. Each of these individual variables could be summarized using a measure of central tendency. The goal, of course, is to better understand the survey respondents—to tell us what these students are like. The following section discusses three measures of central tendency: the mode, the median, and the mean.

The Mode

The mode is the value of the most frequent observation. The mode can be used to help understand data measured at any level. To obtain the mode, you count the number of observations in each distinct category of a variable or the most frequently observed score. The mode is the category of the variable that contains the most observations. Let's say you have a variable named "fruit" and the categories are 1 = mangoes, 2 = apples, and 3 = oranges. Furthermore, you have a basket filled with these fruits; you count the fruits and find that there are 23 mangoes, 15 apples, and 19 oranges in the basket. The mode is "1," which is the value that you assigned to mangoes.

SPSS can quickly find the mode of any variable by producing a frequency table. In addition to the mode, SPSS's frequency table method also provides a count and then calculates percentages found in each of the variable's categories. Using this approach, one can reduce many thousands of cases into a format that can be better understood. In our current work, we use the **class_survey2** database that was originally entered in Chapter 6.

- Start SPSS, and click **Cancel** in the SPSS Statistics opening window.
- Click **File**, select **Open**, and click **Data**.
- In the file list, locate and click **class_survey2.sav**.
- Click **Analyze**, select **Descriptive Statistics**, and then click **Frequencies**.
- Click **Student's Predicted Grade** and **Gender** while holding down the computer's **Ctrl** key.
- Click the arrow to move the selected variables to the Variable(s): panel.
- Click **OK**.

Once **OK** is clicked, the Output Viewer opens, which displays a table titled "Student's Predicted Final Grade" (see Figure 11.1) and another titled "Gender" (see Figure 11.2).

In our quest to better understand the data, let's begin by looking at the frequency table shown in Figure 11.1. There are four categories, and we find that the same number of students predicted grades of "B" and "C." The identification of the mode informs us that there are two modes—we have a bimodal distribution for the variable. How does the frequency table add to our understanding of the data for this single variable? One thing we can say is that there were 24 students (32.4% + 32.4% = 64.8%) who believed that they would earn a grade of "B" or "C." There was not a single student who predicted a grade of "F," which was one of the survey options. Furthermore, we may say that the students appeared to be rather optimistic in that almost 30% predicted a grade of "A."

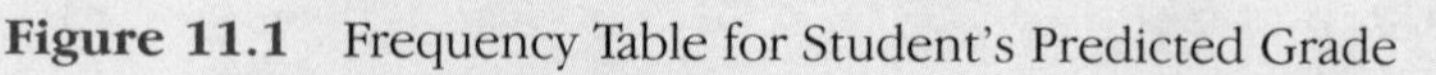

Figure 11.1 Frequency Table for Student's Predicted Grade

Student's Predicted Grade

		Frequency	Percent		
Valid	A	11	29.7	29.7	29.7
	B	12	32.4	32.4	62.2
	C	12	32.4	32.4	94.6
	D	2	5.4	[illegible]	100.0
	Total	37	100.0	[illegible]	

The next variable, "Gender" (see Figure 11.2), indicates that 26 of the 37 students (70.3%) were female—the mode. At this point in our analysis, we could say that we had a class of mostly female students who were rather optimistic about receiving an excellent grade in the statistics class. Note that further analysis would be required to determine whether gender was related to predicted grade. Remember that our descriptive analysis, up to this point, only describes one variable at a time.

Figure 11.2 Frequency Table for Gender

Gender

		Frequency	Percent		
Valid	male	11	29.7	29.7	29.7
	female	26	70.3	70.3	100.0
	Total	37	100.0	100.0	

Next, we examine the median as a measure of central tendency.

The Median

The median is the value that occupies the middle point of a distribution. It is the point that divides the distribution in half. Half the values fall at or below the median and half fall at or above it. We next demonstrate the use of SPSS to determine the median.

Assuming that **class_survey2** is still open, use the sequence given below to obtain the median for selected variables from this database.

- Click **Analyze**, select **Descriptive Statistics**, and then click **Frequencies**.
- Click **Points on Exam One** and **Points on Exam Two** while holding down the computer's **Ctrl** key.
- Click the arrow to move the selected variables to the Variable(s): panel.
- Click **Statistics**, click **Median**, and click **Continue**.
- Click **OK** (Figure 11.3 shows the Output Viewer).

Figure 11.3 Statistics Table for the Median

Statistics

		Points on Exam One	Points on Exam Two
N	Valid	37	37
	Missing	0	0
Median		67.00	81.00

The table reports that the median value for Points on Exam One is 67, whereas the median for Points on Exam Two is 81. With this information, we can say that half the students scored less than or equal to 67 points on the first exam; on the second exam, the median was 81 for an improvement of 14 points.

The Mean

Another common name for the mean is the average. The simplest way to find the mean of a distribution is to add all the values and divide this sum by the number of values. You will find that sometimes the mean is used with categorical variables, such as ordinal data. However, this practice is mathematically unsound. As was discussed in Chapter 4, on levels of measurement, it is difficult, or impossible, to demonstrate equal intervals with ordinal data. The current section considers numerical data (SPSS's scale data) the only legitimate use of this descriptive statistic. The mean, as another measure of central tendency, has as its purpose the making of a distribution of values more understandable. Once again, we assume that the **class_survey2** database is open before following the procedure below.

- Click **Analyze**, select **Descriptive Statistics**, and then click **Frequencies**.
- Click **Points on Exam One**, **Points on Exam Two**, and **Average Points On Exams** while holding down the computer's **Ctrl** key.

- Click the arrow to move the selected variables to the Variable(s): panel.
- Click **Statistics**, click **Mean**, **Median**, and **Mode**, and then click **Continue**.
- Click **OK** (Figure 11.4 shows the Output Viewer).

Figure 11.4 Statistics Table for the Mean, Median, and Mode

Statistics

		Points on Exam One	Points on Exam Two	Average Points (based on exams 1 and 2)
N	Valid	37	37	37
	Missing	0	0	0
Mean		65.95	78.11	72.03
Median		67.00	81.00	73.00
Mode		50	93[a]	57[a]

a. Multiple modes exist. The smallest value is shown

As shown in Figure 11.4, the means for these three exams are 65.95, 78.11, and 72.03. If the purpose of descriptive statistics is to make the data more understandable, then how do these means accomplish this? For one thing, when looking at the raw scores for the 37 students, it is impossible to compare the performance on the two tests in any meaningful way. Using the mean shows we can easily see that there was a 12-point improvement on the second test.

△ 11.3 MEASURES OF DISPERSION

There are three basic measures of dispersion discussed next: range, standard deviation, and variance. The range can be calculated on ordinal or scale data, whereas the standard deviation and variance are calculated only with variables measured at the scale level. Perhaps the easiest way to think of these measures of dispersion is that they tell us how "spread out" the values are. With scale data, we are usually concerned with how the data are dispersed around the mean. One of the most important concepts in descriptive statistics is that to accurately describe scale data, you must calculate both a measure of dispersion (standard deviation is best) and a suitable measure of central tendency—most often the mean. If this is not done, then an incomplete and potentially misleading view of the values may result.

We now use SPSS to do the hard work and calculate the measures of dispersion. We also calculate several other descriptive statistics that describe the shape of the distribution. The descriptive statistics (other than the dispersion measures) will also be explained. It is necessary to use a slightly different command sequence than used for the measures of central tendency—so pay careful attention. Make sure that the **class_survey2** database is open, and then follow the procedure given next.

- Click **Analyze**, select **Descriptive Statistics**, and then click **Descriptives**.
- Click **Points on Exam One**, **Points on Exam Two**, and **Average Points On Exams** while holding down the computer's **Ctrl** key.
- Click the arrow to move the selected variables to the Variable(s): panel.
- Click **Options** (the "Descriptives: Options" window opens; see Figure 11.5).
- Click **Mean**, **Std. deviation**, **Variance**, **Range**, **Minimum**, **Maximum**, **S.E. mean**, **Kurtosis**, and **Skewness**, and then click **Continue** (see Figure 11.5).
- Click **OK** (Figure 11.6 appears in the Output Viewer).

Figure 11.5 Options Window

Descriptives: Options

Mean | Sum

Dispersion
Std. deviation | Minimum
Variance | Maximum
Range | S.E. mean

Distribution
Kurtosis | Skewness

Display Order
Variable list
Alphabetic
Ascending means
Descending means

Continue | Cancel | Help

The Output Viewer opens with the requested statistics (the checked boxes in Figure 11.5) as shown in Figure 11.6. The table provides statistics for the three variables measured at the scale level. Let's describe the values contained in this output and how they qualify as descriptive statistics. We must always keep in mind that the intention of the information in the table in Figure 11.6 is to help us better understand these distributions of scores.

Figure 11.6 Descriptive Statistics Table

Descriptive Statistics

	N	Range	Minimum	Maximum	Mean		Std. Deviation	Variance	Skewness		Kurtosis	
	Statistic	Statistic	Statistic	Statistic	Statistic	Std. Error	Statistic	Statistic	Statistic	Std. Error	Statistic	Std. Error
Points on Exam One	37	86	14	100	65.95	4.266	25.951	673.441	-.280	.388	-1.124	.759
Points on Exam Two	37	77	23	100	78.11	2.721	16.551	273.932	-1.021	.388	1.936	.759
Average Points (based on exams 1 and 2)	37	76	24	100	72.03	3.084	18.758	351.874	-.451	.388	-.291	.759
Valid N (listwise)	37											

We have 37 values (scores) for each of the three variables (two tests and the average for those two tests). The range for each is given as 86, 77, and 76. The range is the difference between the maximum and minimum values and as such measures the spread of the test scores.

The mean of the test scores for Exam One is 65.95 with a Std. Error (Standard Error) of 4.266. The Std. Error indicates how well the sample mean of 65.95 estimates the unknown population mean. Details of the meaning of the Std. Error can be obtained in any introductory statistics book.

In addition to the Range (as described above), we have two more measures of variability for our test scores: The first is the Variance and the second is the Std. Deviation. The variance is simply the average of the squared deviations from the mean. For Exam One, the variance is 673.441, whereas the square root of the variance is 25.951—the standard deviation. This represents the average deviation from the mean of 65.95 when all individual scores are taken into account. The standard deviation is most useful when comparing multiple distributions. Therefore, let's compare our standard deviations for Points on Exam One and Exam Two—25.951 and 16.551.

If all the grades were the same on the first test, the standard deviation would be zero. As the variability of test scores increases, so does the size of the standard deviation. The standard deviation of 25.951 indicates that there was some deviation from the mean. The range of 86 also provides evidence that the scores were spread out (from 14 to 100) over the possible values. Comparing Exam One with the distribution of scores for the Points on Exam Two, we see that the test resulted in less variability, having a standard deviation of 16.551 and a range of 77 (23–100). The data clearly indicate that the scores on the second exam were not as spread out as those on the first exam. Whether the differences in the standard deviations are statistically significant or not and why there are differences are not subjects for this textbook on SPSS.

The Shape of the Distribution (Skewness)

Before leaving the descriptive statistics given in Figure 11.6, we briefly examine skewness and kurtosis. Skewness is a way to describe the shape of the distribution in relationship to the normal curve. The normal curve represents a symmetrical distribution of values. It is a curve that results in exactly the same proportion of area under the curve on both sides of the mean. The mean, median, and mode are equal in the normal distribution; see Figure 11.7(b). Data that are approximately normally distributed, as seen in Figure 11.7(b), make it possible to estimate proportions of values by using the normal curve as a mathematical model for your data.

When raw data deviate from the normal distribution, we have a "skewed" distribution. The distribution could have a negative skew as shown in Figure 11.7(a). A negative skew means that the majority of the values tend to be at the high end of the *x*-axis, which results in the median being greater than the mean and a more representative measure of central tendency. If we look at the skewness number for our "Exam One" data in Figure 11.6, we find a minus 0.280 with a Std. Error of 0.388. The minus sign indicates that the distribution of test scores has a negative skew similar to that depicted in Figure 11.7(a).

The skewness numbers for the three test score variables given in Figure 11.6 all indicate minus skews (–0.280, –1.021, and –0.451). You might wish to confirm the fact that the mean is less than the median by looking at Figure 11.4. A positive skew, as shown in Figure 11.7(c), would indicate that the test scores were at the lower end of the *x*-axis with a mean greater than the median. This was not the case for the variables in the class survey.

Figure 11.7 Three Common Distribution Shapes for Skewness

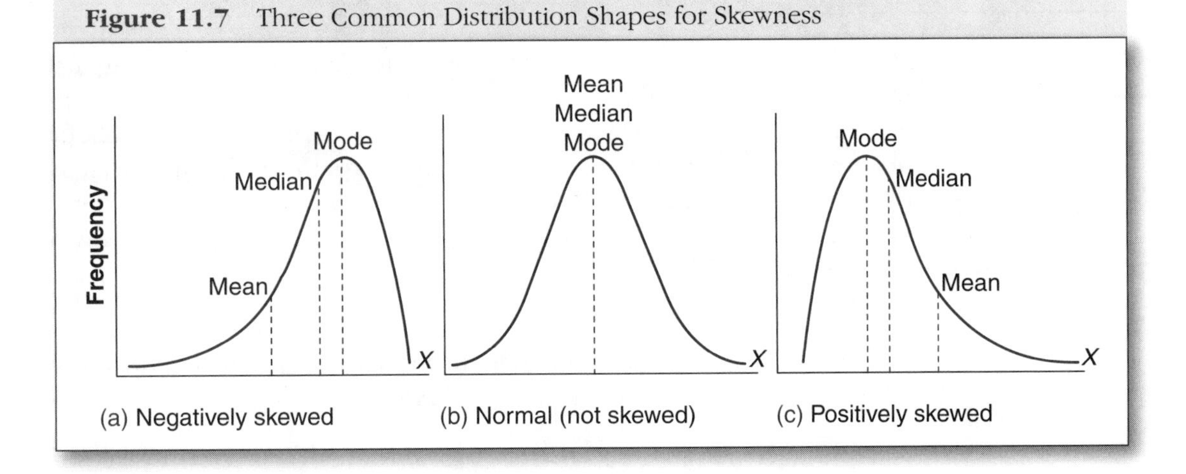

The Shape of the Distribution (Kurtosis)

Another descriptive statistic used to describe the shape of the distribution is kurtosis. As with the skewness, the kurtosis value can be negative, zero, or positive. Figure 11.8(a) shows what a distribution of values would look like if it had negative kurtosis. Basically, it indicates that there is an abundance of cases in the tails—giving it the flat top appearance. Its formal name is platykurtic. Figure 11.8(b) shows the normal curve, whereas Figure 11.8(c) shows a positive kurtosis. A minimum of cases in the tails result in a positive kurtosis, which is known as a leptokurtic distribution.

Figure 11.8 Three Common Distribution Shapes for Kurtosis

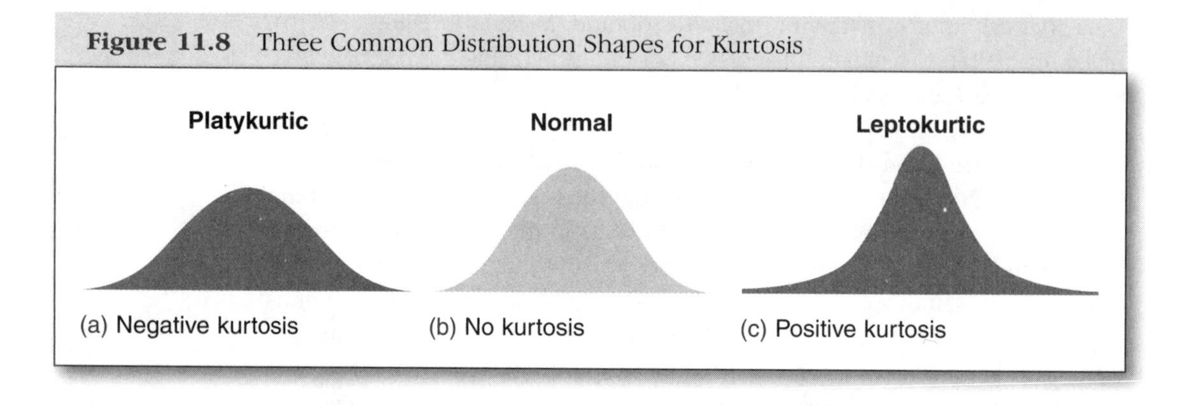

For now, it is sufficient that the reader understands that these descriptive statistics tell us how far a distribution of values deviates from the well-proportioned normal curve.

△ 11.4 The Big Question: Are the Data Normally Distributed?

Many of the inferential statistical tests presented in future chapters require that the data be normally distributed. Therefore, the question of normality is of major importance when making inferences (informed guesses) about unknown population values. There are many ways to answer the question of normality; some involve the use of kurtosis and skewness, whereas some use the power of the SPSS program. Following the procedure below will demonstrate methods used to answer the important normality statistical question.

- Start SPSS, and click **Cancel** in the SPSS Statistics opening window.
- Click **File**, select **Open**, and click **Data**.

- In the file list, locate and click **class_survey2.sav**.
- Click **Analyze**, select **Descriptive Statistics**, and then click **P-P Plots**.
- Click **Average Points On Exams**, and then click the arrow to move this variable to the Variables: panel.
- Click **OK** (when the Output Viewer opens, direct your attention to the P-P Plot as shown in Figure 11.9).

Figure 11.9 Normal P-P Plot for Average Points on Exam

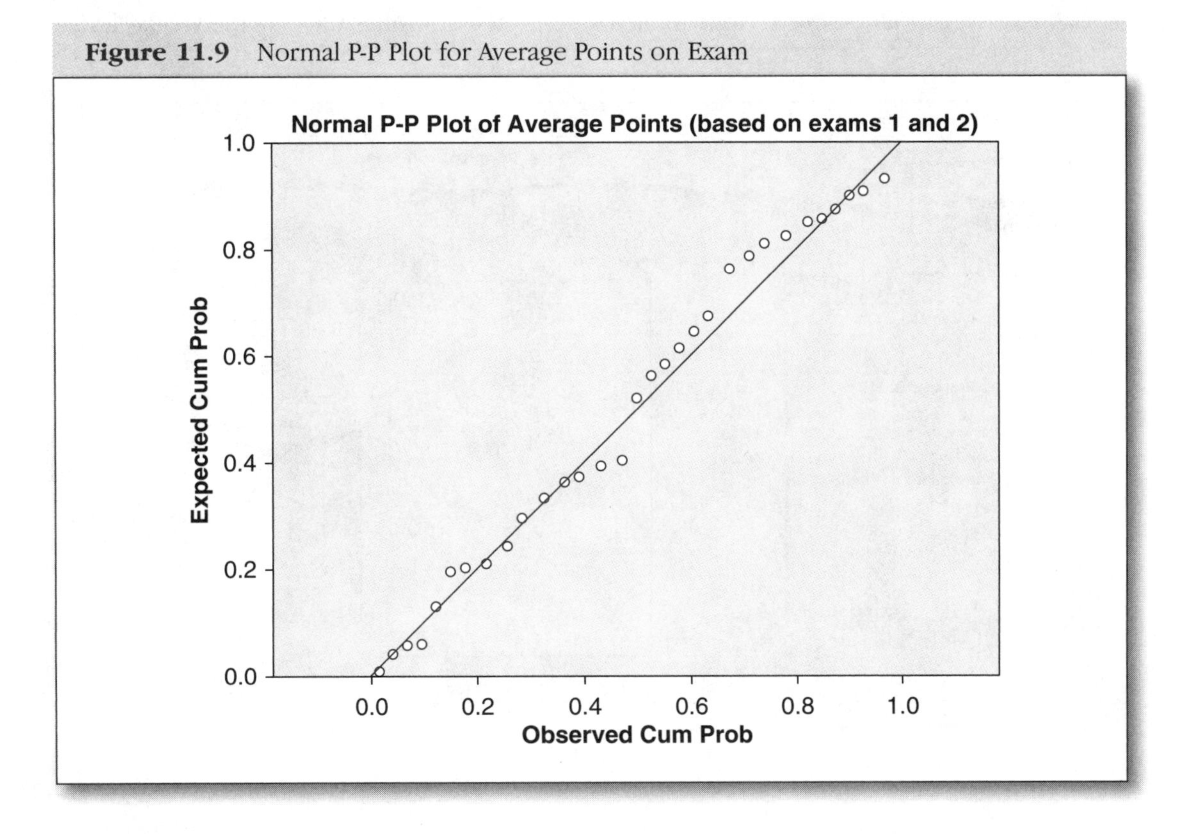

Visual inspection of the P-P Plot would tell us that there is an excellent possibility that the data may be normally distributed. The closer the points are to the line, the greater the probability of normality. Rather than rely on subjective judgment regarding normality, we will next perform a nonparametric test for normality.

- Click **Analyze**, select **Nonparametric**, and click **One-Sample** (the "One-Sample Nonparametric Tests" window opens—not shown).
- Click the **Fields** tab.

- Click **use custom field assignments** (the "One-Sample Nonparametric Tests" window opens).
- Click **Average Points**, and then click the arrow (moves this variable to the Test Fields panel) (see Figure 11.10 for the window's appearance following this click).
- Click **Run** (the Output Viewer displays Figure 11.11).

Figure 11.10 "One-Sample Nonparametric Tests" Window After Clicking the Fields Tab

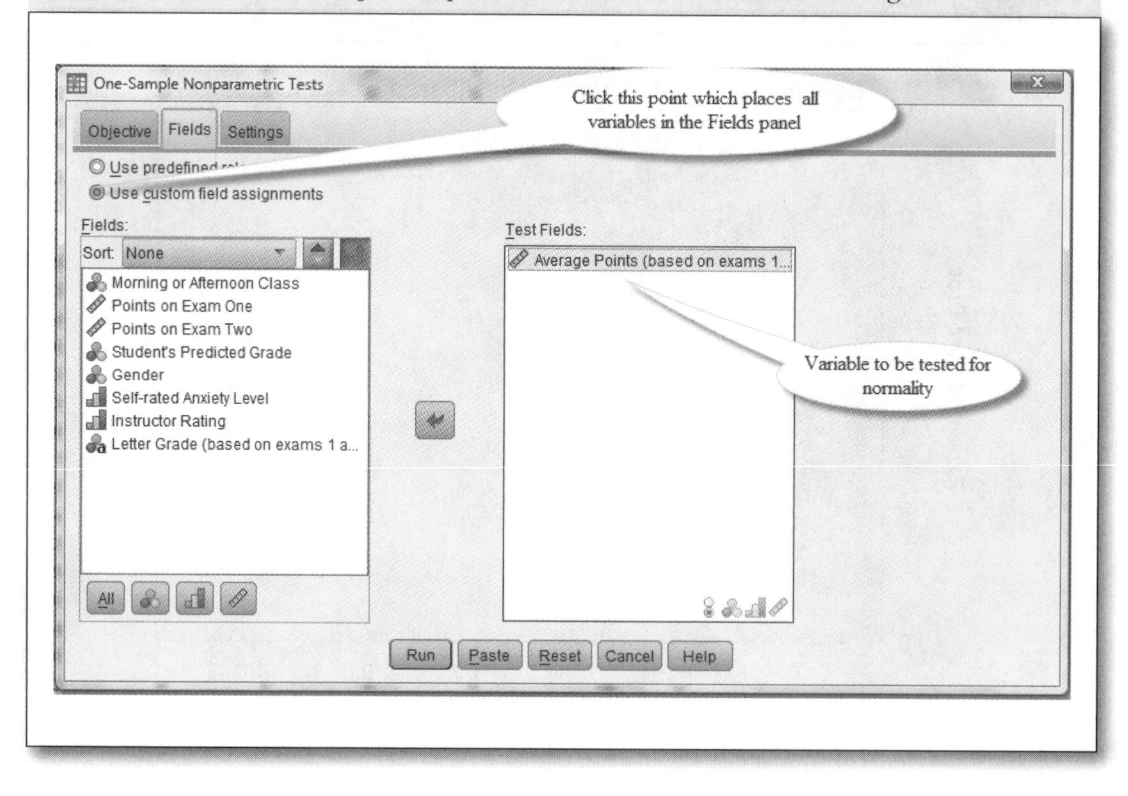

The Output Viewer opens, showing the results of the one-sample Kolmogorov-Smirnov test as indicated in Figure 11.11. The purpose of the Kolmogorov-Smirnov test is to determine if the distribution of values approximates the normal curve. We now have additional evidence that the data are normally distributed as the null hypothesis of normality is not rejected. Note: More information on the concept of testing the null hypothesis can be found in Appendix C.

Figure 11.11 One-Sample Test for Normality: Average Points on Exams

Hypothesis Test Summary

	Null Hypothesis	Test	Sig.	Decision
1	The distribution of Average Points (based on exams 1 and 2) is normal with mean 72.027 and standard deviation 18.758.	One-Sample Kolmogorov-Smirnov Test	.712	Retain the null hypothesis.

Asymptotic significances are displayed. The significance level is .05.

Such questions regarding the distribution's shape are very important since many of the tests in future chapters require that the data approximate the shape of the normal curve.

11.5 DESCRIPTIVE STATISTICS FOR THE CLASS SURVEY △

This chapter opened by indicating that we would show you how SPSS can help make data more understandable. We went on to show how such knowledge would make it possible to discover patterns in the data and assist one in making informed decisions. The data used in the demonstration resulted from a class survey of intermediate statistics students. Let's summarize the SPSS findings that were generated and see if the findings make the data more understandable. What can we say about these students as a result of our descriptive statistical analysis of the data?

We had a rather optimistic group of students, as 62% anticipated either an "A" or "B" in this traditionally challenging class. There were also an equal number of students anticipating a grade of "B" and "C" (32% for both these grades). It was also discovered that 70% of the students were female. We might speculate that the students may have become more serious upon taking the second test as the median score increased by 14 points—from 67 to 81. However, it might also be that the second test was less challenging. Both of these ideas, which account for test score differences, are pure conjecture. Such conjecture is permissible at this stage of the analysis, as offering plausible explanations for any observed pattern can be part of the descriptive process. The mean score also followed the same pattern as the median in that there was a 12-point increase from the first to second exam. The fact that the mean and median values were so close gave us a hint that the data's distribution followed the normal curve.

Measures of dispersion also supported the idea of more serious attitudes toward the material (or an easier second exam) as the variability of test scores declined for the second exam, from 26 to 17, a change of 9 points. It was also shown that the distributions were similar to a normal curve and perhaps subject to further standard statistical tests investigating these differences. These initial observations of normality were confirmed by using the P-P Plots function of SPSS. The nonparametric one-sample test (Kolmogorov-Smirnov test) provided additional evidence that these data approximated the shape of the normal curve.

△ 11.6 SUMMARY

In this chapter, several basic descriptive statistical procedures were described. These procedures were introduced with the idea of summarizing variables so that the analyst could make sense of the data and derive useful information. Descriptive statistical techniques appropriate for variables measured at the nominal, ordinal, and scale (interval and ratio) levels were described. Testing procedures to determine the shape of any distribution, such as whether the data approximate the normal curve, were also described. The following chapter introduces the concept of hypothesis testing and its relationship to inferential statistical analysis.

CHAPTER 12

ONE-SAMPLE *t* TEST AND A BINOMIAL TEST OF EQUALITY

12.1 INTRODUCTION AND OBJECTIVES △

This is the first time we directly address the inferential statistical technique known as hypothesis testing. You may recall that in the previous chapter we presented the results of a hypothesis test for normality with a very brief explanation. This chapter will give many more details, and you will learn about the use of specific hypothesis testing procedures when using the one-sample *t* test and the binomial test of equality. We caution the reader that the concepts of hypothesis testing and their relationship to SPSS presented here must be understood, as they are used in future chapters.

Prior to presenting the purposes of the one-sample and binominal tests, we digress slightly to discuss hypothesis testing and its relationship to inferential statistics. The result of the hypothesis testing procedure is that we discover evidence (or do not) that supports an assertion about population value. Such evidence is derived from data obtained from samples of that population. The reader should realize that our intention in this textbook is not to impart a total understanding of the research process. However, we do intend to give you the knowledge required to select the correct hypothesis test needed to answer the question at hand (various assertions about population values).

Let's describe the purpose of the one-sample *t* test and binomial test of equality. The purpose of the one-sample *t* test is to test whether a sample mean is significantly different from some hypothesized value in a population. The purpose of our binomial test of equality is to determine whether 50% of the values fall above and below the hypothesized population value.

OBJECTIVES

After completing this chapter, you will be able to

- Describe data requirements for using the one-sample *t* test
- Write a research question and null hypothesis for the one-sample *t* test
- Conduct and interpret the one-sample *t* test using SPSS
- Describe when it is appropriate to use the binomial test of equality
- Conduct and interpret the binomial test as an alternative to the one-sample *t* test

12.2 Research Scenario and Test Selection

A review of several different studies conducted in Southern California resulted in the belief that jack rabbits, sucessfully crossing a busy highway, travel at an average speed of 8.3 miles per hour (mph). A conservation officer in Northern California had the idea that the jack rabbits of Northern California travel at a significantly different speed. Armed with a radar gun, the officer positioned himself by the side of Interstate 80 somewhere north of Lake Tahoe. His task was to record the speed of jack rabbits as they crossed the busy highway. Twenty jack rabbits, selected at random, were clocked, and their average speed was determined to be 8.7 mph. What is the appropriate statistical test to answer this conservation officer's question?

The reasoning leading to the selection of the correct test that will answer the officer's question follows. The speeds of the Southern and Northern California jack rabbits were measured at the scale level in miles per hour. Since scale data are a requirement of all *t* tests, it is therefore possible to use one of the *t* tests. It is clear that the conservation officer has only one random sample—how about the one-sample *t* test? The mean rabbit speed of this sample could then be compared with the hypothesized speed of 8.3 mph, which was based on the Southern California rabbits.

Also, the various *t* tests require that the original populations be approximately normally distributed and that they have about the same variability. In the current situation, the assumption of normality and equal variances is based on prior research. The normality requirement is lenient as the *t* test is quite insensitive to non-normal data. Statisticians describe this quality by saying that the test is robust. Given that the data were measured at the scale level and we can justify the assumptions of normality and equal variances for both populations, the one-sample *t* test is selected.

12.3 RESEARCH QUESTION AND NULL HYPOTHESIS

The research question is the researcher's idea, or, in other words, the reason for doing the research. In the world of scientific inquiry, the research question is referred to as the alternative hypothesis and designated as H_A. As mentioned above, the researcher's idea is that the average speed of Northern California jack rabbits is different from the average speed of Southern California rabbits of 8.3 mph. The alternative hypothesis is written in a statistical format as follows: H_A: $\mu \neq 8.3$. This simply states that the mean of the Northern rabbits does not equal 8.3 mph. We are attempting to develop statistical evidence that will support the alternative hypothesis.

The null hypothesis always states the opposite of the researcher's idea (H_A). Therefore, the null hypothesis, designated as H_0, states that the average speed of Northern rabbits *is* 8.3 mph. The null hypothesis written in a statistical format is H_0: $\mu = 8.3$, which indicates that the mean speed of the Northern rabbits equals the speed of the Southern rabbits. In most cases, and indeed in this instance, the researcher wishes to reject the null hypothesis (H_0), which would then provide evidence in support of the alternative hypothesis (H_A).

12.4 DATA INPUT, ANALYSIS, AND INTERPRETATION OF OUTPUT

Let's begin by setting up a new SPSS database that records the speeds of 20 jack rabbits crossing a busy section of Interstate 80 in Northern California. The speeds (in miles per hour) are as follows: 9.53, 7.50, 6.21, 8.95, 10.35, 6.30, 5.20, 12.51, 6.35, 10.23, 9.56, 6.57, 11.78, 10.56, 7.24, 6.19, 10.86, 7.25, 8.34, and 12.78.

- Start SPSS.
- Click **Cancel** in the SPSS Statistics opening window.
- Click **File**, select **New**, and click **Data**.
- Click **Variable View**.
- Click the cell in Row 1 and Column 1 and type **rabbits**.
- Click the cell in the Label column and type **Jack Rabbit Speeds** (select 2 decimals and scale as the level of measurement).

- Click **Data View**, and in the rabbits column type the rabbit speeds of **9.53**, **7.50**, and so forth through **12.78** (you should have 20 rows of data once you complete this data entry).
- Click **File**, click **Save As**, and then type **rabbits** in the file name box.
- Click **Save**.

Now that you have entered the data and saved the file, let's answer the researcher's question using the power of SPSS to do the required calculations. Assuming that the **rabbits.sav** database is open, do the following.

- Click **Analyze**, select **Compare Means**, and then click **One-Sample T-Test** (the "One-Sample T Test" window opens).
- Click **Jack Rabbit Speed**, and then click the arrow.
- Click **Test Value:** box, type **8.3** (following these operations, the One-Sample T Test window should appear as seen in Figure 12.1).

Figure 12.1 "One-Sample T Test" Window

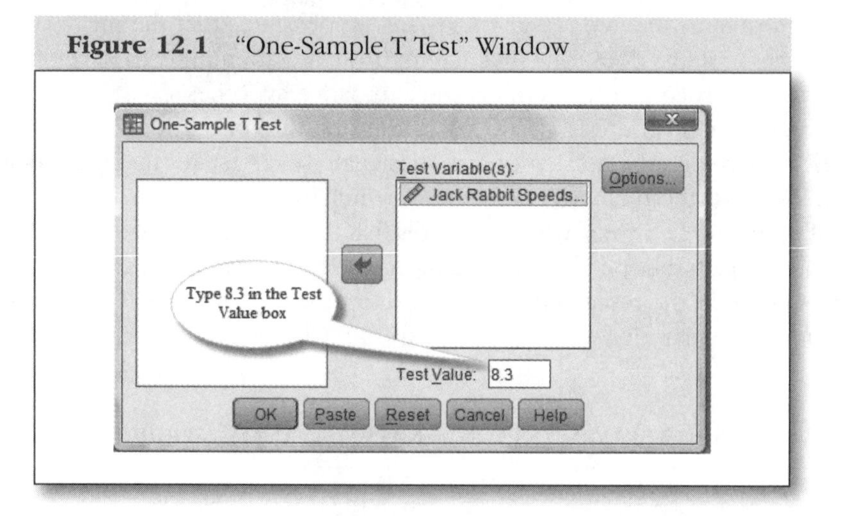

- Click **OK** (the Output Viewer opens; see Figures 12.2 and 12.3).

Figure 12.2 summarizes the descriptive statistics (mean, standard deviation, and standard error) for the rabbits variable.

Figure 12.2 One-Sample Statistics for Northern California Rabbits

One-Sample Statistics

	N	Mean	Std. Deviation	Std. Error Mean
Jack Rabbit Speeds	20	8.7130	2.31356	.51733

The most important aspect of this rabbit research scenario is the answer to the research question: The researcher's idea (and the basis for the research question) is that the speed of Northern California jack rabbits is different from the average speed of Southern California jack rabbits of 8.3 mph. One value in Figure 12.3 answers this important question: It is .435, found in the column titled "Sig. (2-tailed)." "Sig." is an abbreviation for "Significance."

Figure 12.3 One-Sample *t* Test for Jack Rabbit Speeds

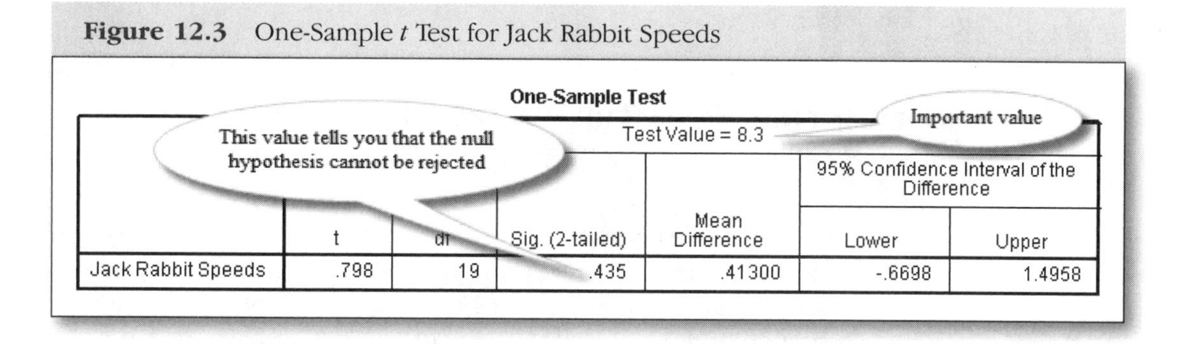

One-Sample Test

	Test Value = 8.3					
					95% Confidence Interval of the Difference	
	t	df	Sig. (2-tailed)	Mean Difference	Lower	Upper
Jack Rabbit Speeds	.798	19	.435	.41300	-.6698	1.4958

Let's explain how the value of .435 answers the research question. The value of .435 directly refers to the H_0 (null hypothesis) and gives us the information required to decide whether the null hypothesis should be rejected or not. Since .435 is greater than .05, we *fail* to reject the null hypothesis, which stated that the average speeds of the Northern and Southern rabbits are the same. Remember that we can only generate support for our research idea if the data indicate that we can reject the null hypothesis (H_0: $\mu = 8.3$). In this case, we were unable to accomplish this.

It is critical that you recognize that the number in the "Sig." column must be small (less than .01 or .05 in most cases) before the null hypothesis can be rejected. The reason for this is that the value of .435 represents the probability that the observed difference is due to chance only. Another way to look at this is to multiply the probability by 100 ($100 \times .435 = 43.5\%$) and recognize that there is a 43.5% chance that the difference was the result of random movement of the data. In this case, we conclude that the average observed speed for Northern rabbits of 8.7 mph is not significantly different from the 8.3 mph recorded for the Southern rabbits. Our conclusion is that the differences are attributable to random movements in the data. In other words, we say that the null hypothesis remains in force—the average speed of Northern California rabbits is 8.3 mph (H_0: $\mu = 8.3$).

As a way to further explain the concept of small values in the "Sig." column, let's assume that .003 was displayed in the "Sig." column. The value of .003 could only have occurred if our sample data resulted in an average speed for the Northern California rabbits that was much different from 8.3 mph. In this case, the value of .003 (a small number) would indicate that there was only a .003 probability that the observed difference was attributable to chance. Therefore, we could reject the null hypothesis and would have evidence in support of the alternative hypothesis that the speeds were significantly different.

To summarize, we say that small numbers in the "Sig." column indicate that detected differences are due to something other than chance (perhaps rabbit characteristics), whereas large numbers indicate that the differences are due to chance.

△ 12.5 NONPARAMETRIC TEST: THE BINOMIAL TEST OF EQUALITY

The binomial test of equality can be used as an alternative if the assumptions of normality and equal variability are not met. It should be noted that the binominal test is principally used with categorical data to test the equality of proportions. However, the jack rabbit scenario offers a viable opportunity for the application of the binomial test of equality. The reasoning is that if the speed of the Northern rabbits is approximately the same as that of Southern rabbits, we would expect an equal number of speeds to fall above and below this value. If the unknown speed is not the same, then the proportion of cases above and below would be significantly different. The reader might also note that the binomial test requires that the scale data be transformed into categorical data; this is done seamlessly by SPSS.

The alternative hypothesis when using the nonparametric binomial test of equality is that the proportion of Northern jack rabbits running above and below the average speed of the Southern rabbits (8.3 mph) will *not* be equal. The null hypothesis is that this proportion will be equal. Let's conduct our binomial test on our jack rabbit speed data and then compare the results with the findings of the parametric test.

- Start SPSS.
- Click **Cancel** in the SPSS Statistics opening window.
- Click **File**, select **Open**, and click **Data**.
- In the file list, locate and click **rabbits.sav**, and then click **Open**.
- Click **Analyze**, select **Nonparametric Tests**, select **Legacy Dialogs**, and then click **Binomial** (the "Binomial Test" window opens; see Figure 12.4).
- Click **Jack Rabbit Speeds**, and then click the arrow.
- Click **Cut Point** and type **8.3**.
- Check the Test proportion box making sure it reads 0.50 (see Figure 12.4).
- Click **OK** (the Output Viewer opens; see Figure 12.5).

Figure 12.4 "Binomial Test" Window for Jack Rabbit Data

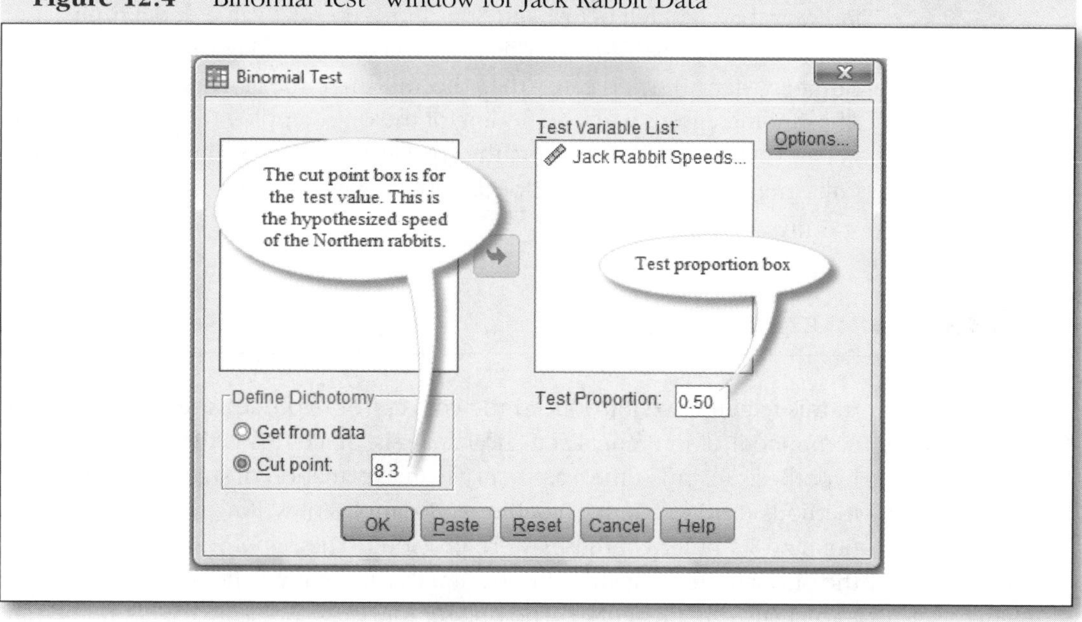

Figure 12.5 Binomial Test Results: Output for Jack Rabbit Data

Binomial Test

		Category	N	Observed Prop.	Test Prop.	Exact Sig. (2-tailed)
Jack Rabbit Speeds	Group 1	<= 8.3	9	.45	.50	.824
	Group 2	> 8.3	11	.55		
	Total		20	1.00		

Figure 12.5 shows that the binomial test procedure separates the jack rabbit speeds into two groups (look at the column titled Category): Group 1 consisted of speeds less than or equal to 8.3 mph and Group 2 consisted of speeds greater than 8.3. The reasoning is that the proportions of observations for the Northern California rabbits (separated into two groups) should be the same (50% in each category). This would be true if the mean speed of the Northern California rabbits was approximately 8.3 mph. If the mean speed of our Northern California rabbits was significantly different, then the proportions would not be equal in the two groups. For instance, let's say the Northern California rabbits sprinted across the highway at the blistering

average speed of 24 mph. In this case, the binomial would be unlikely to indicate that 50% fell below and 50% above the test value of 8.3 mph.

The "Exact Sig. (2-tailed)" is reported as .824 in Figure 12.5. This value (greater than .05) indicates that the null hypothesis cannot be rejected. This finding agrees with the finding of the one-sample *t* test. We have insufficient evidence to support the researcher's idea that the Northern California and Southern California rabbits have significantly different highway crossing speeds.

△ 12.6 SUMMARY

In this chapter, we introduced the concept of hypothesis testing as a major component of inferential statistical analysis. Since this is the first chapter on hypothesis testing, time was spent on an explanation of steps involved in the methods used to conduct such tests. Examples of writing the research question and its null hypotheses were provided. The hypothesis tests known as the one-sample *t* test and a nonparametric alternative, the binomial, were the particular tests presented. The one-sample *t* test basically relies on one random sample of a population, calculates the mean of that sample, and examines whether it equals some hypothesized value. The binomial separates all values into two groups, and then uses the hypothesized value to test whether the observed values are equally distributed above and below the hypothesized value. If they are not so distributed, we have some evidence that the test value (hypothesized value) may be some other value.

Chapter 13

Independent-Samples *t* Test and Mann-Whitney *U* Test

13.1 Introduction and Objectives △

This chapter continues the theme of hypothesis testing as an inferential statistical procedure. In the previous chapter, we investigated whether there was a significant difference between the mean of one random sample of a population and some hypothesized mean for that population.

We now address the situation where we compare two sample means. The test is known as the independent-samples *t* test, and its purpose is to see if we can develop statistical evidence that the two population means are significantly different. You may recall that the sample means are estimates of the unknown means in the sampled population. Therefore, if a significant difference is detected in the sample means, we make an inference that the unknown means of the population are also different. In the next section, you are given details regarding the independent nature of the samples and how the samples may be obtained for the independent-samples *t* test. You are also given the data assumptions required to use this test.

The Mann-Whitney *U* test is the alternative nonparametric test that may be used when the data assumptions required of the independent-samples *t* test cannot be met. Rather than comparing means, which requires scale data, it uses the ranks of the values. Using ranks only requires that the data be measured at the ordinal level. However, the ultimate purpose of the Mann-Whitney *U* test is the same as the independent-samples *t* test—to provide statistical evidence that the sampled populations are significantly different.

OBJECTIVES

After completing this chapter, you will be able to

- Describe data assumptions appropriate for using the independent-samples *t* test
- Write the research question and null hypothesis for the independent-samples *t* test
- Input data for, conduct, and interpret the independent-samples *t* test using SPSS
- Describe circumstances appropriate for use of the Mann-Whitney *U* test
- Conduct and interpret the Mann-Whitney *U* test

△ 13.2 RESEARCH SCENARIO AND TEST SELECTION

The scenario involves an investigation meant to determine if two makes of automobiles obtain significantly different gas mileages. The dependent variable is the number of miles per gallon (mpg) for the Solarbird and the Ecohawk. Recall that the dependent variable is the variable that is subject to change as a result of the manipulation of the independent variable. In this example, the independent variable is the type of automobile, Solarbird and Ecohawk. The experiment will use 12 Solarbird and 12 Ecohawk automobiles, each driven over identical courses for 350 miles each. What would be the appropriate statistical test to determine if the Solarbird and the Ecohawk get significantly different average gas mileages?

We know that miles per gallon is a scale datum, which is a requirement of the *t* test. We also understand that the test vehicles will be randomly selected from a wide range of dealerships in the western United States. Random selection is another requirement for the *t* test. Prior research has also shown that the values for miles per gallon follow a normal curve and that the variances are approximately equal. All data requirements for the *t* test have been met.

Only one question remains—will the samples be independent? As the scenario is explained, there will be two samples taken from two independent populations of Solarbirds and Ecohawks. Thus, we will have two independent samples. Based on this information, we select the independent-samples *t* test for this investigation.

Before moving on to Section 13.3, we address how samples in the independent-samples *t* test may be obtained. Often the concern over how the two samples may be obtained is a source of confusion when attempting to select the appropriate *t* test.

In the scenario just presented, it is very clear that you have two populations; a random sample is obtained from each, and then the sample means are compared. However, the clarity that you have two independent samples is not always the case. For example, another sampling method might require that you take two random samples from one population and then compare the means of the two samples. Yet another alternative would be to take one random sample and divide this sample into two groups, perhaps males and females, and compare the means of these two groups. Regardless of the sampling process, the major consideration, for the independent samples *t* test, is that the measurements taken on the samples must be independent. Independence means that the measurement is taken on another individual or object (e.g., an automobile). This is in contrast with the hypothesis test (paired-samples *t* test) presented in Chapter 14, in which the measurements are taken on the same individual or object but at different times.

13.3 RESEARCH QUESTION AND NULL HYPOTHESIS △

Before reading the next sentence, it would be instructive for you to look away and just visualize the researcher's question or the reason for conducting the investigation. The researcher's idea is that there are statistically significant differences in the average (mean) miles per gallon for the Solarbird and Ecohawk automobiles. For the purpose of the testing procedure, we refer to the researcher's idea as the alternative hypothesis, and we write it as H_A: $\mu_1 - \mu_2 \neq 0$. In plain language, the expression simply states that the difference between the population means (for all Solarbirds and Ecohawks) is not equal to zero. The alternative hypothesis agrees with the researcher's idea—that there are differences in average miles per gallon for the Solarbird and Ecohawk automobiles.

The null hypothesis states the opposite and is written as H_0: $\mu_1 - \mu_2 = 0$. Once again in plain language, the null hypothesis depicts the outcome that the difference between the average miles per gallon for the two populations of automobiles is equal to zero. Remember that if the null hypothesis is rejected, the researcher will have statistical evidence in support of the alternative hypothesis.

△ 13.4 DATA INPUT, ANALYSIS, AND INTERPRETATION OF OUTPUT

We begin by entering the miles per gallon (mpg) data for the 24 vehicles participating in this investigation. Solarbird and Ecohawk miles per gallon data are presented in Figure 13.1.

Figure 13.1 Miles per Gallon Data for the Independent-Samples *t* Test

Solarbird	34.5	36.2	33.2	37.0	32.7	33.1	30.5	37.2	33.5	32.0	36.2	35.7
Ecohawk	38.5	39.2	33.2	39.0	36.7	35.1	38.7	36.3	33.5	34.9	36.8	37.7

Next, we set up the SPSS database and then let the SPSS program do the hard work of analysis. The new database will consist of two variables: one for the mpg data for all 24 vehicles and another "grouping" variable. The grouping (SPSS's term) variable simply labels the mpg data as coming from a Solarbird or Ecohawk.

- Start SPSS.
- Click **Cancel** in the SPSS Statistics opening window.
- Click **File**, select **New**, and click **Data**.
- Click **Variable View**, type **mpg** (name of first variable), and type **miles per gallon** under the Label column (select 2 decimals and scale measure).
- Remain in Variable View and type **make** (name of second variable) and type **make of car** under the Values column (set decimals to zero and specify the nominal measure).
- Click the cell under Values (the "Values" window opens) and type **1** in the Value box and **Solarbird** in the Label box, and then click **Add**. Type **2** in the Value box and type **Ecohawk** in the label box, click **Add**, and then click **OK**.
- Click **Data View** and type in all mpg data beginning with **34.50** and ending with **37.70** (you should have data for all 24 cars entered in one column).
- Click the first cell below the "make" variable and type **1** in the first 12 rows and then type **2** in the next 12 rows.
- Click **File**, click **Save As**, and then type **miles_per_gallon** (note the underscores in the file name).
- Click **Save**.

You have now entered and saved the miles per gallon data for all vehicles. The data entry method just described, where you have all values for the dependent variable (miles per gallon) in one column, is required before SPSS will perform the test. Some may think it would be more logical to enter two variables, one for Solarbird's mpg and one for Ecohawk's mpg, but SPSS does not work that way. Now let's do the fun part—the analysis—and see if we can discover the answer to the research question.

- Click **Analyze**, select **Compare Means**, and then click **Independent-Samples T Test**.
- Click **miles per gallon**, and then click the upper arrow (moves the test variable to the right panel).
- Click **make of car**, and then click the lower arrow (moves the Grouping Variable to the right panel, and the window should look like Figure 13.2) (you may notice the question marks in the Grouping Variable box—don't worry as these will go away once you define the groups in the next step).

Figure 13.2 "Independent-Samples T Test" Window

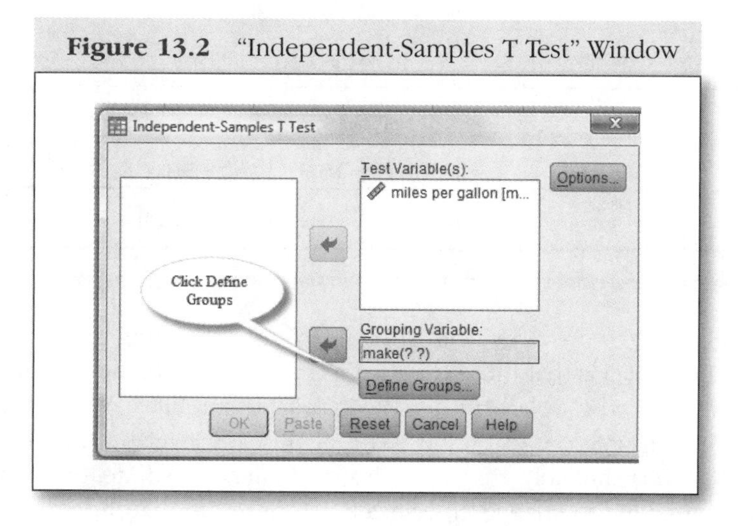

- Click **Define Groups** (the Define Group window opens; see Figure 13.3).

Figure 13.3 "Define Groups" Window

- Click the **Group 1** box and type **1** (1 is the Value for the Solarbird).
- Click the **Group 2** box and type **2** (2 is the Value for the Ecohawk) (completing this and the prior bullet point will eliminate the two Grouping Variable question marks seen in Figure 13.2).
- Click **Continue**, and then click **OK** (the Output Viewer opens; see Figures 13.4 and 13.5).

The Output Viewer opens, displaying the tables shown in Figures 13.4 and 13.5. Figure 13.4 summarizes the descriptive statistics (mean, standard deviation, and standard error) for the miles per gallon variable. At first glance, we see that the Solarbird attained 34.3167 average mpg, whereas the Ecohawk managed an average mpg of 36.6333. The standard deviations for average mpg are close to being identical (2.14257 and 2.09299).

Figure 13.4 Group Statistics for Miles per Gallon

Group Statistics

	make of car	N	Mean	Std. Deviation	Std. Error Mean
miles per gallon	Solarbird	12	34.3167	2.14257	.61851
	Ecohawk	12	36.6333	2.09299	.60419

Figure 13.5 provides the answer to the research question. The researcher's idea is that there is a significant difference in the average miles per gallon for the Solarbird and Ecohawk automobiles. What does the independent-samples t test tell us? The significance level of .014 informs us that it is very unlikely that the observed mean absolute difference of 2.31667 was due to chance. Specifically, we can state that there was a .014 probability that the observed difference was the result of chance and that the null hypothesis can be rejected. The difference can be taken seriously,

Figure 13.5 Independent-Samples Test for Miles per Gallon

Independent Samples Test

		Levene's Test for Equality of Variances		t-test for Equality of Means						
									95% Confidence Interval of the Difference	
			Sig.	t	df	Sig. (2-tailed)	Mean Difference	Std. Error Difference	Lower	Upper
miles per gallon	Equal variances assumed	.080	.780	-2.679	22	.014	-2.31667	.86464	-4.10982	-.52351
	Equal variances not assumed			-2.679	21.988	.014	-2.31667	.86464	-4.10988	-.52346

Large number (greater than .05) indicates equal variances

Small numbers (less than .05) and we can reject the null hypothesis and there are significant differences in average miles per gallon attained by the Solarbird and Ecohawk. We further conclude that the Ecohawk gets superior gas mileage (36.63 mpg) when compared with the Solarbird's average of 34.32 mpg.

From the researcher's standpoint, we can say that the investigation was a success. The null hypothesis of no difference was rejected, and the researcher now has statistical evidence in support of the idea that these two makes of automobiles have different rates of gas consumption.

13.5 NONPARAMETRIC TEST: MANN-WHITNEY *U* TEST △

The miles per gallon data for the Solarbirds and Ecohawks were found to meet all the assumptions (scale data, equal variances, and normally distributed) required for the independent-samples *t* test. However, we wish to demonstrate the SPSS test when these assumptions are not met. We will use the same database for this demonstration. We expect that the less sensitive Mann-Whitney *U* test will also provide evidence that the gas mileage for the two vehicle makes will be significantly different.

The Mann-Whitney *U* test is the nonparametric test selected as the alternative to the independent-samples *t* test. The Mann-Whitney *U* test uses data measured at the ordinal level. Thus, SPSS ranks the miles per gallon scale data and then performs the statistical operations. The observations from both groups are combined and ranked, with the average rank assigned in the case of ties. If the populations are identical in location, the ranks for miles per gallon should be randomly mixed between the two samples.

The alternative hypothesis is that the distributions of miles per gallon (ranks) are not equally distributed between the Solarbirds and Ecohawks. The null hypothesis is that the distributions of miles per gallon (ranks) are equal for both the Solarbirds and the Ecohawks.

Follow the bullet points to use SPSS's Mann-Whitney *U* test to discover if the null hypothesis can be rejected.

- Open SPSS and open **miles_per_gallon.sav**.
- Click **Analyze**, select **Nonparametric Tests**, and then click **Independent Samples**.
- Click **Fields** in the "Nonparametric Tests: Two or More Independent Samples" window.
- Click **miles per gallon**, and then click the upper arrow.
- Click **make of car**, and then click the lower arrow.
- Click **Run** (the "Output Viewer" window opens; see Figure 13.6).

Figure 13.6 Hypothesis Test Summary for Miles per Gallon

Hypothesis Test Summary

	Null Hypothesis	Test	Sig.	Decision
1	The distribution of miles per gallon is the same across categories of make of car.	Independent-Samples Mann-Whitney U Test	.015	Reject the null hypothesis.

Asymptotic significances are displayed. The significance level is .05.

The less sensitive nonparametric test equivalent (Mann-Whitney *U* test) also found a significant difference in the miles per gallon attained by these two automobiles. The null hypothesis of equality was rejected just as in the independent-samples *t* test.

△ 13.6 SUMMARY

This chapter presented the parametric independent-samples *t* test and the nonparametric Mann-Whitney *U* test. Data were analyzed, which compared gas cunsumption (in miles per gallon) for two different makes of automobiles. The research was being conducted because the investigator suspected that the gas consumption would be different for the Solarbirds and Ecohawks. Data were given, and SPSS was used to generate inferential statistics for both parametric and nonparametric tests. Both indicated a significant difference, thus providing statistical evidence in support of the researher's idea. In Chapter 14, the parametric paired-samples *t* test and its nonparametric analog, the Wilcoxon signed-ranks test, are presented and contrasted with the independent-samples tests presented here.

Chapter 14

Paired-Samples *t* Test and Wilcoxon Test

14.1 Introduction and Objectives △

We continue the theme of the previous two chapters by showing how SPSS is used to accomplish hypothesis testing. Chapters 12 and 13 covered the one-sample *t* test and the two-sample independent *t* test. Also addressed were their nonparametric alternatives, the binomial test of equality and the Mann-Whitney *U* test.

This chapter presents the paired-samples *t* test, which compares measurements (means) taken on the same individual or on the same object but at different times. First addressed is the parametric test—the paired-samples *t* test. The nonparametric alternative, the Wilcoxon test, compares ranks for the two measurements and is covered later in the chapter.

The purpose of the paired-samples test is to test for significant differences between the means of two related observations. Usually, the test is used when an individual or object is measured at two different times. Such an investigative approach is often described as a pre- and posttest research methodology. A mean value is determined, based on a pretest, some action intervenes, such as an educational lecture, and another identical measure (the posttest) is taken. Using this methodology, we look for a significant difference between the means of our pre- and posttests. If the investigation was well designed, meaning that all other factors were controlled for, then we can attribute significant differences to the intervening action. In the current example, the intervening action is the educational lecture. A significant difference (hopefully posttest scores were greater) in scores between the means of the pre- and posttests would provide evidence that learning took place as a direct result of the lecture.

The paired-samples t test can be used in many different situations, not just when measuring human performance. One example might be to select a random sample of Harley Davidson motorcycles and record the miles per gallon for brand X gasoline. You could then drain the tanks and refill with brand Y, record miles per gallon, and look for a significant difference between miles per gallon for brand X and Y.

When our distribution of differences fails to meet the criteria of normality, the Wilcoxon signed-ranks test is the nonparametric alternative. The Wilcoxon test demonstrated in this chapter compares two sets of values (scores) that come from the same individual or object. This situation occurs when you wish to investigate any changes in scores over a period of time. The difference between the paired-samples t test and the Wilcoxon test is that the latter does not require normally distributed difference data. The Wilcoxon test is able to use data measured at the ordinal, interval, or ratio level. Therefore, in our current example the interval data (blood pressure levels) could be ranked, and the median ranks for the two categories are then tested for significant differences.

OBJECTIVES

After completing this chapter, you will be able to

- Describe the data assumption requirements when using the paired-samples *t* test
- Write a research question and alternative and null hypotheses for the paired-samples *t* test
- Use SPSS to answer the research question for the paired-samples *t* test
- Describe the circumstances appropriate for using the Wilcoxon nonparametric test
- Use SPSS to conduct the Wilcoxon test
- Interpret SPSS's output for the Wilcoxon test

△ 14.2 Research Scenario and Test Selection

Whenever human subjects are used in research, it is necessary that the principal investigator obtain approval to ensure that the participants are not harmed. We are not sure if the research scenario we are about to describe would pass a review by a Human Subjects Committee, but let's assume that it did pass. The basic concept is that an unexpected violent and chaotic event would significantly change human systolic blood pressure readings. In other words, the researcher intends to measure blood pressure (pretest), scare the heck out of human subjects, and then measure the blood pressure (posttest) once again.

Twelve men aged 20 to 25 years were randomly selected from a group of recruits that had volunteered for service in the United States Marine Corps. Early in their training, they were called to a barracks and their systolic blood pressure was recorded by 12 technicians. The prestimulus measurements are the systolic blood pressure readings. Following the measurement procedure, the Marine Corps recruits were instructed to relax, and during this period of "relaxation" a realistic recording of nearby gunfire, explosions, and yelling was played from loudspeakers just outside the barracks. The simulated chaos is called the "stimulus" in the research method selected for this study. The recruits were commanded to take cover and take no action. The simulated chaos continued for several minutes until the "all clear" was announced. The young Marine Corps recruits were then retested for systolic blood pressure levels by the same 12 technicians. This was the poststimulus measurement. What is the appropriate test for this scenario?

The reasoning process for test selection should start with the level of measurement. Since we have scale data and we are looking for differences between means, you should initially consider one of the *t* tests. Since we have pairs of measurements that were taken on the same subject (pre- and poststimulus on each Marine Corps recruit), we select the paired-samples *t* test. For the paired-samples test, the differences between the pairs of observations are assumed to be distributed normally. The larger the sample size, the more likely the assumption of normality will be met. In spite of our small sample ($n = 12$) in the Marine Corps example, we assume normality and proceed with the test.

14.3 RESEARCH QUESTION AND NULL HYPOTHESIS △

As you may have already guessed, the researcher's idea is that the pre- and poststimulus blood pressure readings will be significantly different. Stated in statistical terms, we write the alternative hypothesis as H_A: $\mu_1 - \mu_2 \neq 0$. This expression states that the difference between the population means for our two sample measurements is not equal to zero. If you discover that the means are not equal, then the data agree with the researcher's idea.

You now have evidence of significant differences in blood pressure readings pre- and poststimulus.

Note: The reader may notice that the researcher's idea would most likely be that there would be an increase in blood pressure readings following the simulated violent chaos. This is technically known as a "directional" hypothesis. In most cases, when using SPSS, it is unnecessary to specify directionality. If significance is found, we can simply look at whether there was an increase or decrease in blood pressure. We are then justified in stating that there is either a significant increase or decrease in blood pressure following the simulated chaos. Such directionality is illustrated when the analysis of our data is completed and SPSS's output is interpreted.

The null hypothesis states the opposite and is written as H_0: $\mu_1 - \mu_2 = 0$. This expression states that the difference between the population means is equal to zero. The researcher is investigating the scenario to determine if the null hypothesis can be rejected. If the null hypothesis is rejected, then we have some evidence that the alternative hypothesis is true. In this case, this would mean that there is a significant difference in the blood pressure readings.

△ 14.4 DATA INPUT, ANALYSIS, AND INTERPRETATION OF OUTPUT

We begin by entering the two systolic blood pressure readings for the 12 young Marine Corps recruits participating in this experiment. Prestimulus blood pressure (**bl_pres1**) readings are 122, 126, 132, 120, 142, 130, 142, 137, 126, 132, 128, and 129. Poststimulus blood pressure (**bl_pres2**) readings are 133, 136, 136, 132, 137, 130, 146, 142, 123, 137, 134, and 140.

- Start SPSS.
- Click **Cancel** in the SPSS Statistics opening window.
- Click **File**, select **New**, and click **Data**.
- Click **Variable View**, type **bl_pres1** in Column 1, Row 1, and then type **Blood Pressure (pre-stimulus)** below the Label column and set zero decimals and leave scale measure.
- Remain in Variable View and type **bl_pres2** in Column 1, Row 2, and then type **Blood Pressure (post-stimulus)** in the Label column and set zero decimals and leave as a scale measure.
- Click **Data View**, click the cell for Row 1 and Column 1, and type all data for the variable **bl_pres1** beginning with **122** and ending with **129**.
- Click the cell for Row 1 and Column 2 and type all data for the variable **bl_pres2** beginning with **133** and ending with **140**.

- Click **File**, click **Save As**, and then type **blood_pressure** in the File name box of the Save As window.
- Click **Save.**

You have now entered and saved the blood pressure data that were observed for the 12 Marine Corps recruits. We now do the analysis.

- Click **Analyze**, select **Compare Means**, and then click **Paired-Samples T Test** (the "Paired-Samples T Test" window opens; see Figure 14.1).

Figure 14.1 "Paired-Samples T Test" Window: Blood Pressure Variables in the Left Panel

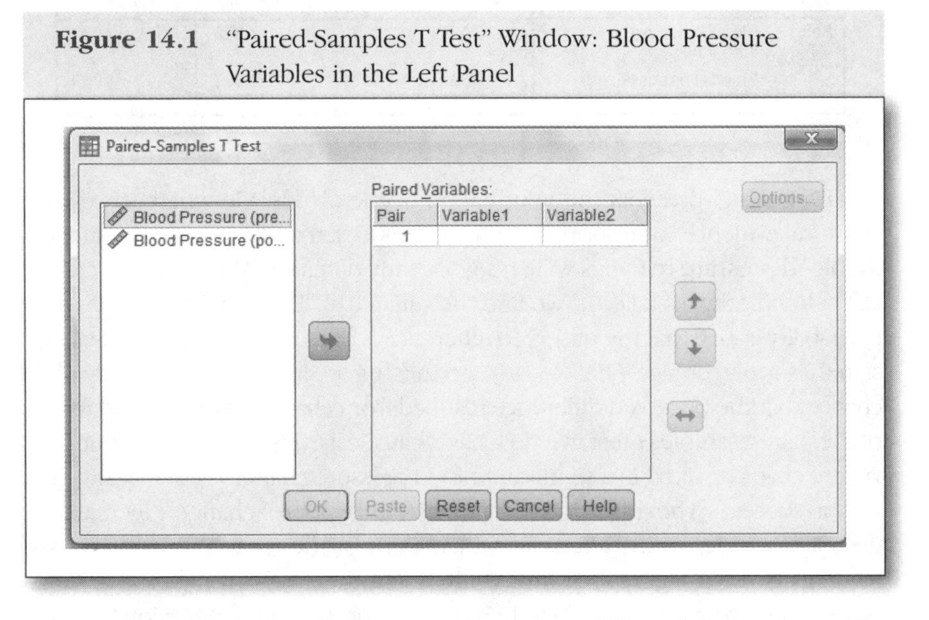

- Click **Blood Pressure (pre-stimulus)**, and then click the arrow (the variable moves to the right panel).
- Click **Blood pressure (post-stimulus)**, and then click the arrow (the variable moves to the right panel).
- Click **OK** (the Output Viewer opens, displaying Figures 14.2 and 14.3).

Looking at Figure 14.2, which presents the basic statistics for each variable, we see that the means for the pre- and posttest are indeed different (130.5 vs. 135.5). The paired-samples *t* test will answer the question whether the observed difference is actually "significant." If our test shows significance, then we can attribute the change in blood pressure to the simulated violent chaos. The reader should note that the investigation was designed to control for other factors that may have influenced the Marine Corps recruits' blood pressure levels. If our test does not find a significant difference, then any change in blood pressure could be attributed to chance. The Std. Deviations (Standard Deviations) for both pre- and posttest measurements are also different. This is not a problem for the paired-samples *t* test. Recall that the assumption for the paired-samples *t* test is only that the differences between pairs be normally distributed.

Figure 14.2 Paired-Samples Statistics: Pre- and Poststimulus Blood Pressure Readings

Paired Samples Statistics

		Mean	N	Std. Deviation	Std. Error Mean
Pair 1	Blood Pressure (pre-stimulus)	130.50	12	7.026	2.028
	Blood Pressure (post-stimulus)	135.50	12	5.916	1.708

Figure 14.3 presents the information needed to decide whether we have statistical evidence to support the researcher's idea that the pre- and poststimulus blood pressure readings will be significantly different. When we look at the last column, "Sig. (2-tailed)," we find the value .010. This value tells us that the probability is .010 that the observed difference is due to chance. The probability, stated as a percentage (100 × .010), permits us to say that there is only a 1% chance that the observed difference resulted from chance. Based on the results of the paired-samples *t* test, we now have statistical evidence that the simulated violent chaos resulted in a significant blood pressure change—the investigation was a success. The direction of the blood pressure change (increase or decrease) can be easily determined by looking at the mean pre- and posttest readings given in Figure 14.2. It is evident that there was an increase in mean blood pressure readings (130.5–135.5) following the simulated violent chaos (stimulus). Since the *t* test found a significant difference, we may now say that there was a significant increase in blood pressure following the simulated chaos.

Figure 14.3 Paired-Samples Test for Blood Pressure Data

Paired Samples Test

		Paired Differences					t	df	Sig. (2-tailed)
		Mean	Std. Deviation	Std. Error Mean	Lower	Upper			
Pair 1	Blood Pressure (pre-stimulus) - Blood Pressure (post-stimulus)	-5.000	5.543	1.600	-8.522	-1.478	-3.125	11	.010

This value (.010) tells us we can reject the Null Hypothesis

Let's now examine the same data with the assumption that it does not meet the requirements of the parametric paired-samples *t* test.

14.5 NONPARAMETRIC TEST: WILCOXON SIGNED-RANKS TEST △

Recall that the paired-samples *t* test assumes that the differences between the pairs are normally distributed. If the differences between the pairs greatly differ from normality, it would be a good idea to use the Wilcoxon signed-ranks test. The Wilcoxon test converts the blood pressure interval data to the ordinal level of measurement. It then calculates the medians of the ranks for the pre- and poststimulus groups, after which it tests to see if there is a significant difference between the medians of the pre- and poststimulus categories.

The alternative and null hypotheses remain the same as with the paired-samples *t* test. The alternative hypothesis states that the difference is *not* equal to zero, whereas the null hypothesis states it does equal zero.

Follow the bullet points to see what the Wilcoxon signed-ranks test informs us regarding the rejection of the null hypothesis.

- Start SPSS.
- Click **File**, select **Open**, and click **Data**.
- Click **blood_pressure.sav**, and then Click **Open**.
- Click **Analyze**, select **Nonparametric Tests**, and then click **Related Samples** (the "Nonparametric Test: Two or More Related Samples" window opens; see Figure 14.4) (note that the window in Figure 14.4

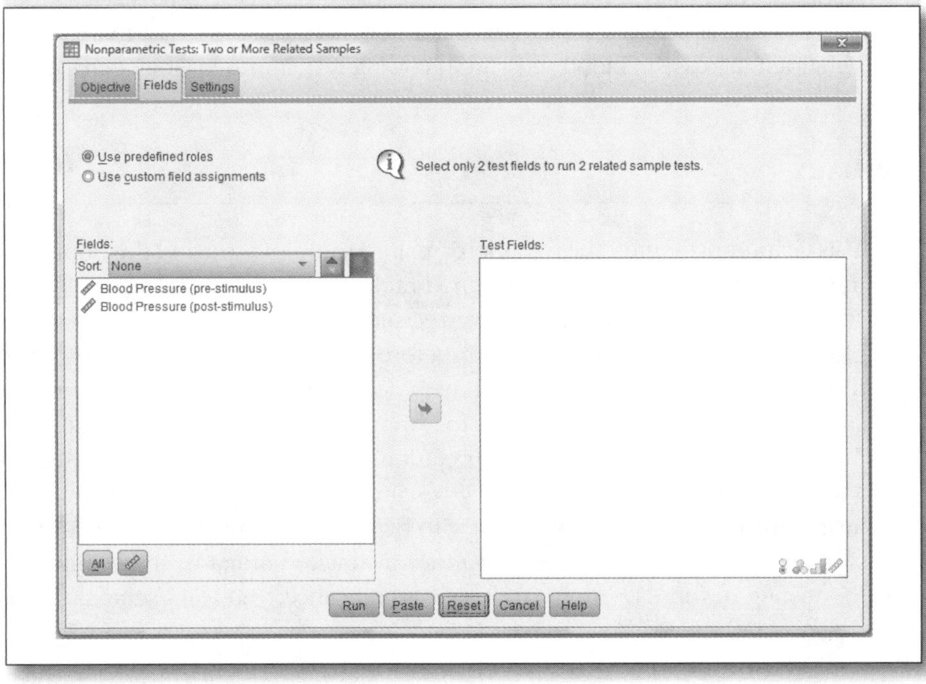

Figure 14.4 "Nonparametric Test: Two or More Related Samples" Window

shows what you see after the Fields tab is clicked—see the next bullet point).

- Click the **Fields** tab at the top of the window if the window does not appear as in Figure 14.4.
- Click **Blood Pressure (pre-stimulus)**, and then click the arrow.
- Click **Blood Pressure (post-stimulus)**, and then click the arrow.
- Click **Run** (the Output Viewer opens; see Figure 14.5).

Figure 14.5 Hypothesis Test: Nonparametric Wilcoxon Signed-Rank Test

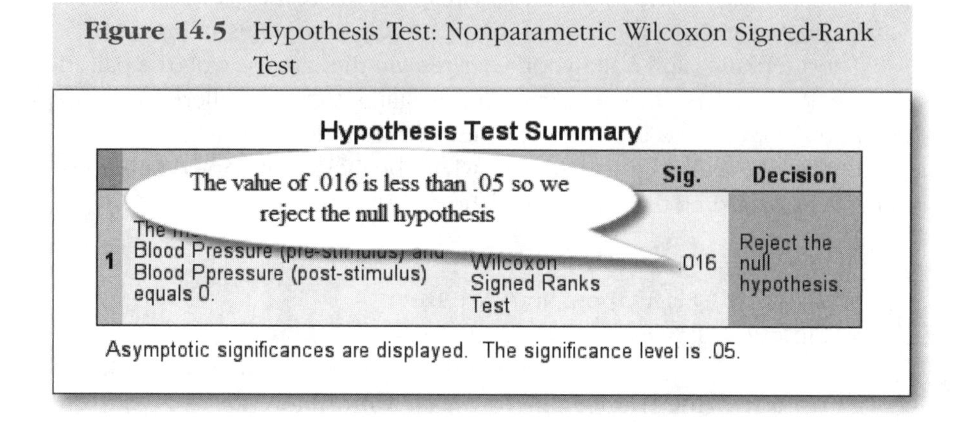

The less sensitive nonparametric test alternative (related-samples Wilcoxon test) also found a significant difference for the pre- and posttest blood pressure readings. The null hypothesis that the difference was equal to zero was rejected, as it was in the paired-samples t test. We now have evidence that there was a significant increase in blood pressure following the simulated violent chaos.

△ 14.6 SUMMARY

This chapter presented the parametric paired-samples t test and the nonparametric alternative, the Wilcoxon signed-ranks test. Data were derived from a study that subjected young military recruits to frightening stimuli. Systolic blood pressure readings were taken before and after the stimulus. Both the tests (t test for parametric data and the Wilcoxon for nonparametric) found a significant change in blood pressure following the introduction of the stimulus. The data input, analysis, and interpretation of SPSS's output were discussed and explained in relationship to the researcher's reason for conducting the investigation. Chapter 15 continues with hypothesis testing but addresses the situation where you have three or more means to compare. To accomplish the testing of three or more means for significance, you will use analysis of variance and the Kruskal-Wallis H test.

Chapter 15

One-Way ANOVA and Kruskal-Wallis Test

15.1 Introduction and Objectives △

We continue with the theme of the prior three chapters in that we use SPSS to accomplish hypothesis testing. Let's list the hypothesis tests covered so far with their nonparametric alternatives given in parentheses: one-sample *t* test (binomial), two-sample independent *t* test (Mann-Whitney), and the paired-samples *t* test (Wilcoxon). You may have noticed that in all these tests we tested various assertions about two means. The hypothesis test presented here also involves the testing of means, the difference being that we now have three or more independent group means that we wish to test for significant differences. The appropriate test for three or more means is the ANOVA, an acronym for analysis of variance. Don't let the name of the test mislead you into thinking it does not represent another case of comparing the means (not the variances as the name may imply) of various groups. It is a procedure that tests for significant differences between three or more means. It determines significance via the calculation of the *F* statistic. The value of *F* is calculated when the variance of the total group is compared with the variances of the individual groups. We refer the reader to any basic statistical text for details on the underlying mathematics used in ANOVA.

With ANOVA you have one independent variable. Recall that the independent variable is thought to have an effect on the dependent variable. The independent variable may be measured at the scale (interval/ratio) or categorical (nominal/ordinal) level. The dependent variable must be measured at the scale level. The one independent variable may consist of any number of groups but at least three. Each group of the independent variable represents a unique treatment. These treatments are most often referred to as "levels" of the independent variable. Variables designated as string (alphabetic nominal) in SPSS are permissible as the independent variable for the ANOVA procedure.

For example, we have one independent variable, antibacterial spray, but we have four different brands of the spray, labeled A, B, C, and D. The levels of the independent variable are A, B, C, and D; they could also be referred to as four different treatment groups. The dependent variable could be the number of bacteria killed by an application of the various levels of the one independent variable, bacterial spray.

It is important to note that a significant *F* statistic, resulting from the ANOVA test, only tells you that there is a significant difference between at least two group means, while not identifying which two are different. To answer which of the pairs are significantly different, we must conduct post hoc analysis. Post hoc simply means "after the fact." When significance is established, then we do additional work (after the fact) to identify which of the multiple pairs of means contributed to the significant *F* statistic. SPSS makes this additional work easy.

As in the previous chapters, we also present a nonparametric analog when data characteristics fail to meet the minimum data requirements for using the ANOVA test. The Kruskal-Wallis test is the nonparametric alternative often used when the level of measurement is ordinal. The Kruskal-Wallis approach is to rank the data and compare the median of the ranks for all groups with the individual group medians. The mathematical procedures are similar to those used in ANOVA, except now we compare medians rather than means. If Kruskal-Wallis identifies overall significance, then SPSS can examine each pair for significance (pairwise comparisons).

OBJECTIVES

After completing this chapter, you will be able to

- Describe data assumptions required for the ANOVA hypothesis test
- Write a research question, alternative hypothesis, and null hypothesis for ANOVA
- Use SPSS to answer the research question for the ANOVA procedure
- Conduct post hoc analysis (Scheffe test) for ANOVA when significance is discovered

Describe data characteristics required for the Kruskal-Wallis nonparametric test

Use SPSS to conduct the Kruskal-Wallis test

Interpret SPSS's output for the Kruskal-Wallis test

15.2 RESEARCH SCENARIO AND TEST SELECTION

The Municipal Forest Service (MFS) had the idea that the amount of monies spent to maintain five disparate mountain hiking trails in the Santa Lucia Mountains were significantly different. Random samples of such expenditures were taken from the records for the past 50 years. Six years were selected at random and expenditures for trail maintenance recorded. The MFS wished to use the findings of their analysis to help them allocate resources for future trail maintenance.

The dependent variable is the amount of money spent on each of the five hiking trails. The independent variable, hiking trails, consists of five separate trails. Since the data were measured at the scale level, we may calculate means and standard deviations for the amount of money spent. The mean expenditures for each trail were based on the random samples of size six ($n = 6$). We assume that the distributions of trail expenditures are approximately normally distributed with equal variances. Which statistical test would you use to develop evidence in support of the MFS's belief that the amount of money spent on trail maintenance was significantly different for the five trails?

We have scale data for our dependent variable (expenditures in dollars); therefore, we are able to look for differences between means. We can eliminate t tests from consideration since we have more than two means. At this point, the ANOVA, designed to compare three or more means, seems like a worthy candidate. To use ANOVA, there must be random selection for sample data—this was accomplished. The distributions of expenditures appear to approximate the normal curve, and their variances approximate equality. Based on this information, we select the ANOVA. If the ANOVA results in significance, the Scheffe post hoc analysis will be used to identify which pairs of means contributed to the significant F value.

15.3 RESEARCH QUESTION AND NULL HYPOTHESIS

The researcher's idea, and reason for conducting this investigation, is the belief that there is a significant difference in maintenance expenditures for the five hiking trails. Recall that the alternative hypothesis (H_A) is simply a restatement of the researcher's idea. We write the following statistical expression: H_A: $\mu_1 \neq \mu_2 \neq \mu_3 \neq \mu_4 \neq \mu_5$. In plain language, the expression simply states that the means of the populations, based on data from the random samples, are not equal.

The null hypothesis (H_0) states the opposite and is written as H_0: $\mu_1 = \mu_2 = \mu_3 = \mu_4 = \mu_5$. The expression H_0 states that there are no differences between the means of the populations. The MFS researcher would prefer to reject the null hypothesis, which would then provide statistical evidence for the idea that maintenance expenditures for the five trails are significantly different.

△ 15.4 DATA INPUT, ANALYSIS, AND INTERPRETATION OF OUTPUT

We begin by setting up a new database and then entering the expenditure data recorded by the MFS. SPSS requires that the expenditure data be entered as one variable and another as a grouping variable. The procedure to accomplish the data entry is provided in the bullets following Figure 15.1. The amount of money, in thousands of dollars, spent to maintain each of the five trails is given in Figure 15.1.

Figure 15.1 Municipal Forest Service Trail Maintenance in Thousands of Dollars (Santa Lucia Mountains)

	Six Randomly Selected Years					
Trail	**Year 1**	**Year 2**	**Year 3**	**Year 4**	**Year 5**	**Year 6**
Eaton Canyon (1)	20	11	17	23	29	19
Grizzly Flat (2)	11	3	8	13	11	12
Rattlesnake (3)	7	5	3	9	8	14
Bailey Canyon (4)	3	3	1	3	8	14
Millard Canyon (5)	16	21	16	11	6	16

Let's next create a new database in a format compatible with the ANOVA procedure.

- Start SPSS.
- Click **Cancel** in the SPSS Statistics opening window.
- Click **File**, select **New**, and click **Data**.
- Click **Variable View**, type **Dollars** in the cell found in Row 1 and Column 1, and then type **Dollars (in thousands) spent on trail maintenance** beneath the Label column, set zero decimals, and leave as scale measure.

- Type **Trail** in the cell found in Row 2 and Column 1, and then type **Name of Trail** in the Label column.
- Click the **cell** in Row 2 in the Values column, and then click the button that is located in that cell (the Value labels window opens).
- Add trail information in the Value Labels window: Begin by typing the value **1** and Label **Eaton Canyon** and click **Add**, and repeat this process for all the five trails where **2** = **Grizzly Flat**, **3** = **Rattlesnake**, **4** = **Bailey Canyon**, and **5** = **Millard Canyon**; when all trail information is entered, click **OK** (set decimals to **zero** and change level of measurement to **Nominal**).
- Click **Data View**, click the cell for Row 1 and Column 1, and type data for Eaton Canyon in Rows 1 through 6: **20**, **11**, **17**, **23**, **29**, and **19**. Repeat this process for the remaining data that are shown in Figure 15.1. Continue entering data in Row 7 and Column 1 for the Grizzly Flat data and continue until all the Dollar variable data are entered. When finished, there should be 30 values in Column 1 of this new database.
- Click the cell beneath the trail variable and type **1** for the six rows of this variable, type **2** for the next six rows, type **3** for the next six rows, type **4** for the next six rows, and finally type **5** for the last six rows of data (see Figure 15.2 for the ANOVA data entry procedure for the first 10 cases).

Figure 15.2 Data View for One-Way ANOVA (First 10 of 25 Cases Shown)

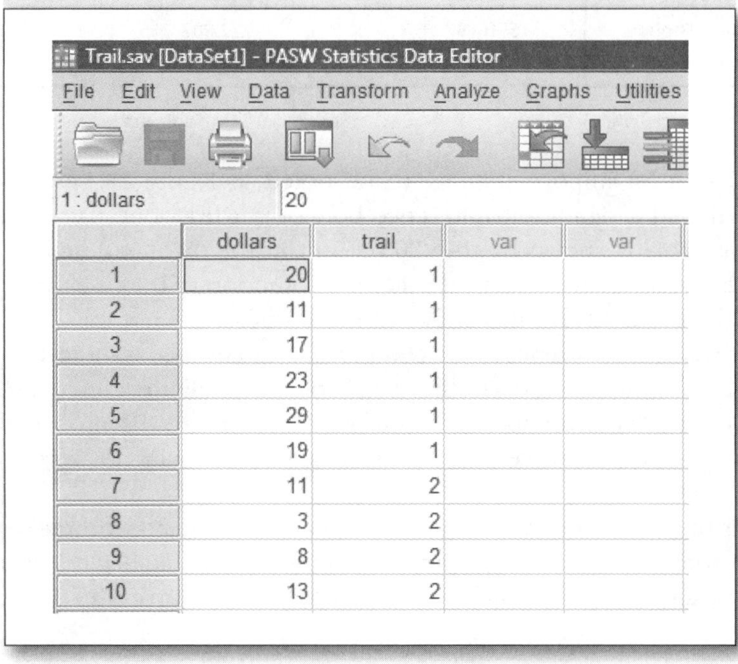

Trail.sav [DataSet1] - PASW Statistics Data Editor

File Edit View Data Transform Analyze Graphs Utilities

1 : dollars 20

	dollars	trail	var	var
1	20	1		
2	11	1		
3	17	1		
4	23	1		
5	29	1		
6	19	1		
7	11	2		
8	3	2		
9	8	2		
10	13	2		

- Click **File**, click **Save As**, and then type **trail**.
- Click **Save**.

You have now entered and saved the MFS expenditure data for the five trails of interest. Let's proceed to the analysis using the power of the SPSS ANOVA procedure.

- Click **Analyze**, select Compare Means, and click **One-way ANOVA**.
- Click **Dollars (in thousands)**, and then click the upper arrow (moves the variable to the Dependent List box).
- Click **Name of Trail**, and then click the lower arrow (moves variable to the Factor box).
- Click **Post Hoc** (the "One-way ANOVA: Post Hoc Multiple Comparisons" window opens).
- Click **Scheffe**, click **Continue**, and then click **OK** (the Output Viewer opens; see Figures 15.3 and 15.4).

Figure 15.3 ANOVA Table for Trail Expenditure

ANOVA

Dollars (in thousands) spent on tr...

The value of .000 permits the rejection of the null hypothesis

	Su... Squar...				Sig.
Between Groups	800.800	4	200.200	8.778	.000
Within Groups	570.167	25	22.807		
Total	1370.967	29			

The most important thing to be noted in Figure 15.3 is that SPSS determined a significant difference between at least one pair of the five means. This is shown by the .000 found in the "Sig." column. In other words, we now have statistical evidence in support of the researcher's idea that the expenditures for all five of these trails were not equal. As in the prior tests for differences between means, significance is indicated by the small value (.000) shown in the "Sig." column in Figure 15.3. (Note: SPSS rounds the "Sig." number to three places; therefore the probability that the observed differences are due to chance is actually not zero but less than .0005.)

If the researcher were doing this analysis by hand, the real work of identifying which trails were significantly different would begin. Fortunately, we clicked **Scheffe,** which instructed SPSS to compare all 10 possible combinations of mean values for these five hiking trails. This comparison is presented in Figure 15.4.

Note: Figure 15.4 may seem a little overwhelming because of its shear length. Be aware that it presents 20 comparisons for two means when there are actually only 10. This is because, for instance, Eaton Canyon is compared with Grizzly Flat and once again Grizzly Flat is compared with Eaton Canyon. The same probabilities are assigned to both comparisons.

Figure 15.4 Post Hoc Analysis: Scheffe Test of Significance (Values Given in Thousands)

Multiple Comparisons

...sands) spent on trail maintenance

(I) Name of Trail	(J) Name of Trail	Mean Difference (I-J)	Std. Error	Sig.	Lower Bound	Upper Bound
Eaton Canyon	Grizzly Flat	10.167*	2.757	.024	[illegible]	[illegible]
	Rattlesnake	12.167*	2.757	.005	[illegible]	[illegible]
	Bailey Canyon	14.500*	2.757	.001	[illegible]	[illegible]
	Millard	5.500	2.757	.429	-3.66	14.66
Grizzly Flat	Eaton Canyon	-10.167*	2.757	.024	[illegible]	-1.01
	Rattlesnake	2.000	2.757	.969	[illegible]	[illegible]
	Bailey Canyon	4.333	2.757	.654	[illegible]	[illegible]
	Millard	-4.667	2.757	.589	-13.83	4.49
Rattlesnake	Eaton Canyon	-12.167*	2.757	.005	[illegible]	-3.01
	Grizzly Flat	-2.000	2.757	.969	[illegible]	[illegible]
	Bailey Canyon	2.333	2.757	.947	[illegible]	[illegible]
	Millard	-6.667	2.757	.244	-15.83	2.49
Bailey Canyon	Eaton Canyon	-14.500*	2.757	.001	[illegible]	-5.34
	Grizzly Flat	-4.333	2.757	.654	[illegible]	[illegible]
	Rattlesnake	-2.333	2.757	.947	[illegible]	6.83
	Millard	-9.000	2.757	.056	-18.16	.16
Millard	Eaton Canyon	-5.500	2.757	.429	-14.66	3.66
	Grizzly Flat	4.667	2.757	.589	-4.49	13.83
	Rattlesnake	6.667	2.757	.244	-2.49	15.83
	Bailey Canyon	9.000	2.757	.056	-.16	18.16

*. The mean difference is significant at the 0.05 level.

The Scheffe test is another test of significance, but this time the test compares each possible combination of means one at a time. Examination of the Scheffe post hoc analysis reveals that there are three mean comparisons that are significantly different. First, we find that there was an average difference of $10,167 per year on the maintenance of the Eaton Canyon and Grizzly Flat trails. Second, it is determined that there is an average difference of $12,167 for expenditures on Eaton Canyon and Rattlesnake. Finally, it is determined that there is an average difference of $14,500 for the Eaton and Bailey Canyon trails. All these differences are statistically significant at the .05 level.

Summarizing our findings, we may state that the maintenance expenditures for these five mountain trails are significantly different. There is now statistical support for the researcher's hypothesis as well as additional details regarding which of the trail pairs are significantly different.

Let's now examine the same data but with the belief that they do not meet the data assumptions required for the ANOVA procedure.

△ 15.5 NONPARAMETRIC TEST: KRUSKAL-WALLIS TEST

The application of the nonparametric alternative for the one-way ANOVA, the Kruskal-Wallis test, is now demonstrated using the same trail maintenance data. The same data are used, but this time let's assume that the data severely differ from the normal distribution.

The Kruskal-Wallis test is similar to ANOVA in many ways. One similarity is that the null hypotheses for parametric and nonparametric tests assume that the random samples are drawn from identical populations. The alternative hypotheses for both ANOVA and Kruskal-Wallis are that the samples come from nonidentical populations. Presented another way, we can state that the alternative hypothesis looks for statistically significant differences between the groups. One thing to be kept in mind is that the Kruskal-Wallis test is not as powerful as ANOVA—it may miss significance.

Assumptions for the use of the Kruskal-Wallis test are (1) that samples are random, (2) that data are at least measured at the ordinal level, and (3) that scaled distributions are identically shaped. Let's next demonstrate the application of the Kruskal-Wallis procedure.

- Start SPSS.
- Click **File**, select **Open**, and click **Data**.
- Click **Trail.sav**, and then Click **Open**.
- Click **Analyze**, select **Nonparametric Tests**, and then click **Independent Samples** (the "Nonparametric Test: Two or More Related Samples" window opens).
- Click **Fields**, click **Dollars**, and then click the upper arrow.
- Click **Name of Trail**, and then click the lower arrow.
- Click **Run** (the Output Viewer opens; see Figure 15.5).

Figure 15.5 Nonparametric Test Summary for Trail Maintenance Costs

Hypothesis Test Summary

	Null Hypothesis	Test	Sig.	Decision
1	The distribution of Dollars (in thousands) spent on trail maintenance is the same across categories of Name of Trail.	Independent-Samples Kruskal-Wallis Test	.003	Reject the null hypothesis.

Asymptotic significances are displayed. The significance level is .05.

The test determines that maintenance costs are significantly different with a low "Sig." value of .003. Recall that this value simply states that the probability that the noted differences are attributable to chance alone is only .003. Since we have rejected the null hypothesis, that dollars spent on the maintenance of each trail is the same, we might wish to do comparisons between all pairs of mountain trails to establish which trail pairs are indeed significantly different. You will recall that a similar procedure was followed when conducting the Scheffe test in the ANOVA procedure.

- Double click on the **Hypothesis Test Summary** in the Output Viewer (see Figure 15.5) (the Model Viewer window opens; see Figure 15.6).

Direct your attention to the lower right portion of Figure 15.6, which is enclosed by the ellipse. The area in the ellipse is enlarged and presented in Figure 15.7.

Figure 15.6 Model Viewer Window: Kruskal-Wallis Test for Trail Maintenance Data

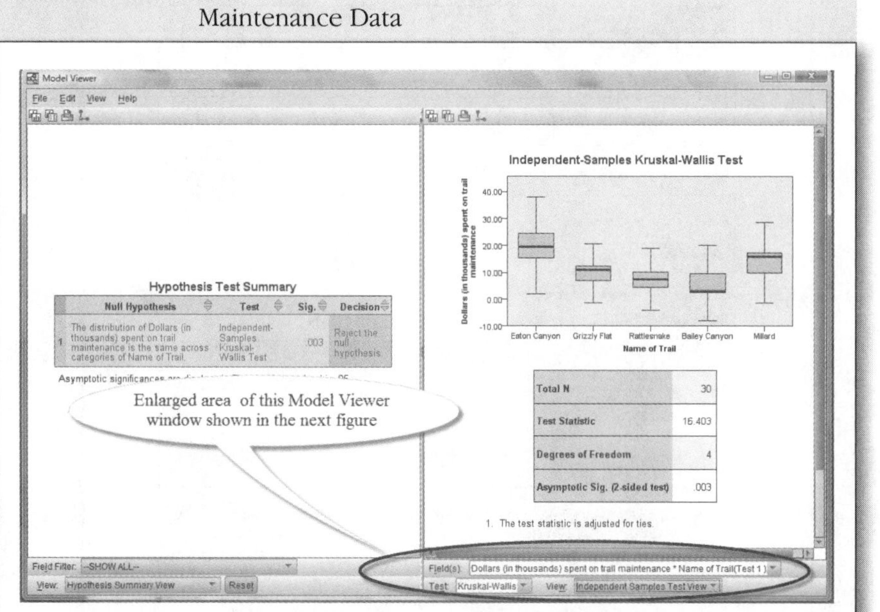

- Click the **arrow** in the View line (this action is shown as "Click here first" in Figure 15.7) (once this is done, the menu, as shown in Figure 15.7, appears).
- Click **Pairwise Comparisons** (see Figure 15.8 for the result of this click).

Figure 15.7 Lower Right Portion of the Model Viewer Window (Enlarged)

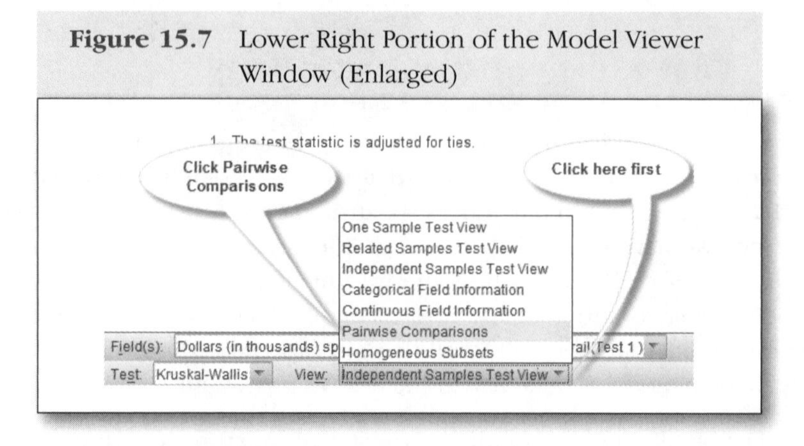

Once **Pairwise Comparisons** is clicked, Figure 15.8 appears in the Output Viewer, which compares all possible pairs of the five mountain hiking trails.

Figure 15.8 Pairwise Comparisons for the Kruskal-Wallis Nonparametric Test

Sample1-Sample2	Test Statistic	Std. Error	Std. Test Statistic	Sig.	Adj.Sig.
Bailey Canyon-Rattlesnake	3.333	5.061	.659	.510	1.000
Bailey Canyon-Grizzly Flat	6.417	5.061	1.268	.205	1.000
Bailey Canyon-Millard	-12.667	5.061	-2.503	.012	.123
Bailey Canyon-Eaton Canyon	18.000	5.061	3.557	.000	.004
Rattlesnake-Grizzly Flat	3.083	5.061	.609	.542	1.000
Rattlesnake-Millard	-9.333	5.061	-1.844	.065	.651
Rattlesnake-Eaton Canyon	14.667	5.061	2.898	.004	.038
Grizzly Flat-Millard	-6.250	5.061	-1.235	.217	1.000
Grizzly Flat-Eaton Canyon	11.583	5.061	2.289	.022	.221
Millard-Eaton Canyon	5.333	5.061	1.054	.292	1.000

Each row tests the null hypothesis that the Sample 1 and Sample 2 distributions are the same.
Asymptotic significances (2-sided tests) are displayed. The significance level is .05.

The Kruskal-Wallis test is less sensitive than the ANOVA. Although both the ANOVA and Kruskal-Wallis tests found significance between Eaton and Bailey and Eaton and Rattlesnake, the nonparametric test did detect differences between Eaton and Grizzly Flat.

15.6 SUMMARY △

In this chapter, we presented the procedures for checking for significant differences between three or more means. The ANOVA was used for data meeting the parametric requirements. The ANOVA requires that data for each of the randomly sampled populations be approximately normally distributed, that they have equal variances, and that the dependent variable be measured at the scale level. Once these conditions are met and the ANOVA test is applied, an F statistic is used to determine if any of the pairs of means are significantly different. If found to be significant, then the data were further analyzed. The purpose of further analysis was to identify those pairs of means that contributed to the significant F statistic. Such post hoc analysis used the calculation of Scheffe's critical value for all possible pairs of means.

When data fail to meet the parametric requirments, the Kruskal-Wallis nonparametric test may be used. This procedure was demonstrated on the same data used to show the ANOVA procedure. The application of Kruskal-Wallis resulted in the determination that there was a significant difference between two of the pairs of means out of the 10 possible comparisons. In the next chapter, we double our fun by addressing the situation where there are two or more independent variables—two-way ANOVA.

Chapter 16

Two-Way (Factorial) ANOVA

△ 16.1 Introduction and Objectives

We continue the theme of the prior three chapters in that we use SPSS to accomplish hypothesis testing. In Chapter 15, you were introduced to one-way analysis of variance (ANOVA), which tested for significant differences between three or more means. Recall that when using one-way ANOVA you have one independent variable that can have three or more groups, called "levels of the independent variable."

This chapter presents the circumstances where you have at least two independent variables with any number of groups (levels) in each. The two-way ANOVA is sometimes referred to as factorial ANOVA. The use of factorial ANOVA permits the evaluation of each independent variable's effect on the dependent variable, which are known as the *main* effects. But it also does more as it tests for *interaction* effects. Interaction means that the factorial analysis looks for significant changes in the dependent variable as a result of two or more of the independent variables working together.

For example, you are interested in investigating how tall corn grows under various conditions—the height of the corn, measured in inches, is the dependent variable. You are interested in studying two independent variables: the first variable is seed type (Pearl and Silver) and the second independent variable is fertilizer type (seaweed and fish). To succinctly describe such a study's design, we state that we are using a 2 × 2 factorial ANOVA. In other words, we have two independent variables; the first has two groups (seed types) and the second also has two groups (fertilizer types).

Let's use the 2 × 2 design and the hypothetical data in Figure 16.1 to explain the main and interaction effects. The main effects investigated would be the seed and fertilizer types. Are there significant differences in the height of corn when grown from Pearl and Silver seeds? Looking at the column means, found in Figure 16.1, we find that the value is 50 inches for both types of seed. Therefore, there is no difference in height, and we conclude that there is no main effect due to seed type. The other main effect would be fertilizer type. Are there significant differences in corn height when the seed is fertilized with seaweed or fish? The data indicate no differences in the height of corn when fertilized with these two fertilizer types. This is shown by the row mean heights of 50 inches for both fertilizer types. However, there is an interaction effect between seed and fertilizer type. Pearl seed corn grows taller when seaweed fertilizer is used (60 inches) rather than fish fertilizer (40 inches). Silver seed corn grows taller when using fish fertilizer (60 inches) rather than with seaweed fertilizer (40 inches).

Figure 16.1 Hypothetical Data for Corn Growing Study (Main and Interaction Effects)

	Pearl Seed (Inches)	Silver Seed (Inches)	Row Means (Inches)
Seaweed Fertilizer	60	40	50
Fish Fertilizer	40	60	50
Column Means	50	50	

The two-way ANOVA is specifically designed to investigate data to determine whether there are main and/or interaction effects.

OBJECTIVES

After completing this chapter, you will be able to

Describe data assumptions required for two-way ANOVA

Input data for the two-way ANOVA procedure

Use SPSS to conduct two-way ANOVA

(Continued)

(Continued)

Describe the main effects in two-way ANOVA

Describe the interaction effects in two-way ANOVA

Write alternative and null hypotheses for two-way ANOVA

Interpret SPSS output to identify main and interaction effects

△ 16.2 Research Scenario and Test Selection

An investigation was proposed that would study two independent variables—detergent type and water temperature—each consisting of two groups; therefore we have a 2 × 2 factorial design. The investigators suspect that the type of detergent and water temperature may have some impact on the whiteness of the clothes (main effects). Those conducting the study also wish to determine if the detergent type and water temperature interact in a manner that may affect the whiteness level (interaction effect).

Thus, we have one dependent variable, whiteness, measured at the scale level and two independent variables. The two independent variables used in this scenario are type of detergent (Mighty Good and Super Max) and water temperature (hot and cold). Both are measured at the nominal level. Since each load of clothing will be washed independently, a repeated measures design is inappropriate. If we begin with the assumption of normality and equal variances of the populations, what statistical test would you suggest for this investigation?

The reasoning for deciding which statistical test is appropriate might start with a consideration of levels of measurement for the selected variables. We have a dependent variable measured at the scale level, which informs us that we can calculate means and standard deviations for our groups. We also have two independent variables measured at the nominal level. Two-way ANOVA procedures will work with nominal independent variables as long as the dependent variable is measured at the scale level. We will use a 2 × 2 factorial design since we have two groups for each of our independent variables. We also assume that the dependent variable is approximately normally distributed and that groups have equal variances. The assumption of normality and equality of variances will be empirically tested as we proceed through the analysis.

16.3 RESEARCH QUESTION AND NULL HYPOTHESIS

The researchers' wish to investigate whether there is a difference in whitening power of Mighty Good and Super Max detergent. For the detergent main effect research question, we have the following alternative hypothesis:

$H1_A$: $\mu_{Mighty\_Good} \neq \mu_{Super\_Max}$

They also wish to investigate if there is a significant main effect on the whiteness of clothes when washed in hot or cold water. The alternative hypothesis for the hot and cold water research question is $H2_A$: $\mu_{hot} \neq \mu_{cold}$.

The null hypotheses state the opposite of the alternative hypotheses. If the data permit the rejection of either of the two following null hypotheses, we can state that there is statistical evidence that there was a significant main effect on the level of whiteness.

$H1_0$: $\mu_{Mighty\_Good} = \mu_{Super\_Max}$

$H2_0$: $\mu_{hot} = \mu_{cold}$

The researchers also wish to determine whether the detergent type and water temperature may work together to change the whiteness of the clothing (interaction effect). Figure 16.6 shows the interaction effects by presenting the mean whiteness levels for (Mighty Good + cold water), (Mighty Good + hot water), (Super Max + cold water) and (Super Max + hot water). Figure 16.8 then reports whether any significant interaction was detected.

In the next section, we input, analyze, and interpret the data from the clothes washing study.

16.4 DATA INPUT, ANALYSIS, AND INTERPRETATION OF OUTPUT

Your new database for this example will be named **whiteness**. We are confident that your skill level is such that detailed instructions regarding the entry of variable information and the data are not required. Figure 16.2 does however provide a representation of what your Variable View page should look like once you have entered all variable information. Note that the Values windows for detergent and water should be completed as follows: values for detergent should be 1 = *Mighty Good* and 2 = *Super Max*, whereas values for water should be 1 = *Hot* and 2 = *Cold*.

Figure 16.2 Variable View Screen for the Whiteness Study

	Name	Type	Width	Decimals	Label	Values	Missing	Columns	Align	Measure
1	detergent	Numeric	8	0	Type of detergent	{1, Mighty Good...	None	8	Right	Nominal
2	water	Numeric	8	0	Water temperature	{1, Hot}...	None	8	Right	Nominal
3	white	Numeric	8	1	Whiteness meter reading	None	None	8	Right	Scale

Enter the variable information, and then carefully enter all data (40 cases) shown in Figure 16.3. Note that you should not enter the data in the column titled "Case #" as SPSS assigns these numbers automatically when each row of data is numbered.

Figure 16.3 Data for the 2 × 2 Factorial Whiteness Study

Case #	Detergent	Water	Whiteness	Case #	Detergent	Water	Whiteness
1	2	1	68.7	21	2	2	65.2
2	2	1	50.1	22	2	2	79.3
3	2	1	58.1	23	2	2	60.2
4	2	1	56.0	24	2	2	67.8
5	2	1	42.3	25	2	2	65.7
6	2	1	52.0	26	2	2	51.9
7	2	1	27.8	27	2	2	61.7
8	2	1	53.4	28	2	2	62.5
9	2	1	56.3	29	2	2	63.1
10	2	1	57.7	30	2	2	66.6
11	1	1	60.9	31	1	2	65.2
12	1	1	61.3	32	1	2	75.9
13	1	1	42.2	33	1	2	57.3
14	1	1	49.8	34	1	2	64.5
15	1	1	47.7	35	1	2	60.9
16	1	1	33.9	36	1	2	50.8
17	1	1	43.7	37	1	2	59.3
18	1	1	19.5	38	1	2	58.4
19	1	1	45.1	39	1	2	58.8
20	1	1	53.6	40	1	2	62.3

The following list presents a much abbreviated procedure for entering the variable information (see Figure 16.2) and the actual data (see Figure 16.3). This is followed by a more detailed explanation of how to get SPSS to do the work required for factorial analysis. Once the Output Viewer displays the results, the various tables are explained.

- Start SPSS.
- Click **Cancel** in the SPSS Statistics opening window.
- Click **File**, select **New**, and click **Data**.
- Click **Variable View** (enter all variable information as represented in Figure 16.2).
- Click **Data View** (carefully enter all data for the three variables as given in Figure 16.3).
- Click **File** and then **Save As**, type **whiteness** in the File Name box, and then click **OK**.
- Click **Analyze**, select **General Linear Model**, and then click **Univariate** (the "Univariate" window opens; see Figure 16.4).
- Click **Whiteness Meter Reading**, and then click the arrow by the Dependent Variable: box.
- Click **Type of Detergent**, and then click the arrow for the Fixed Factor(s): box.
- Click **Water Temperature**, and then click the arrow for Fixed Factor(s): (your window should look like Figure 16.4).

Figure 16.4 "Univariate" Window: All Variables Moved to the Right Panes

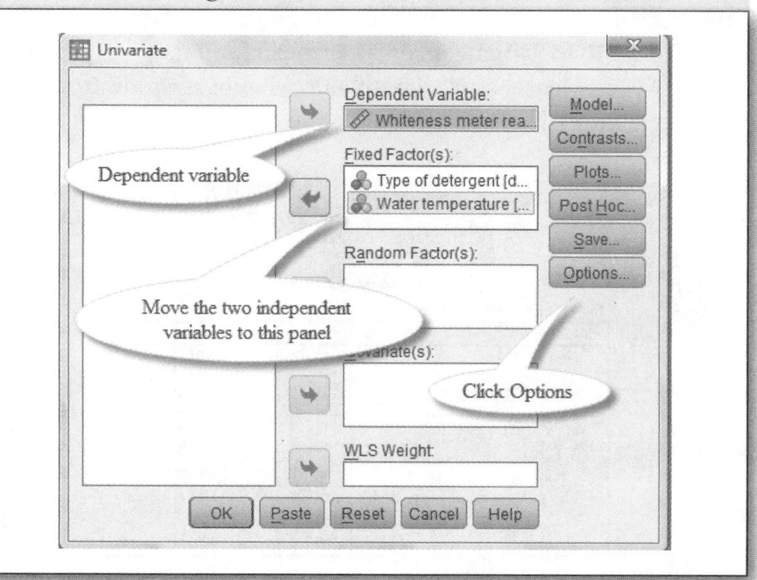

- Click **Options** (the "Univariate: Options" window opens; Figure 16.5).

Figure 16.5 "Univariate: Options" Window

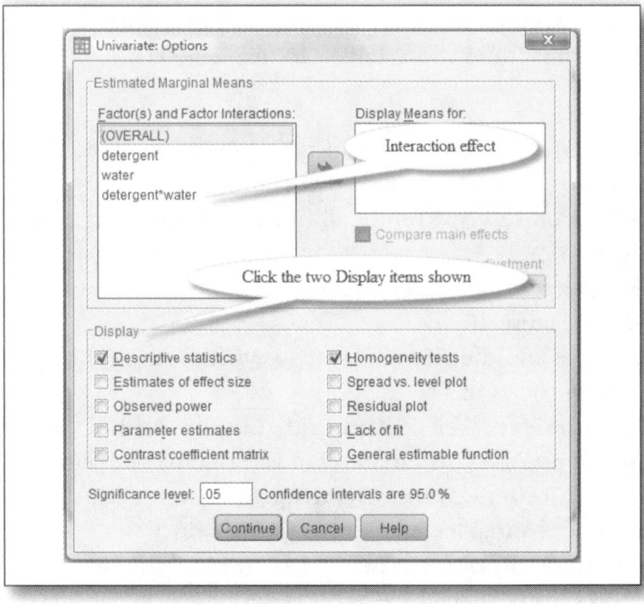

- Click **Descriptive statistics** and **Homogeneity tests**.
- Click **Continue** (the "Univariate" window opens), and then click **OK** (the Output Viewer opens).

The reader is advised that once **OK** is clicked, a considerable amount of SPSS output is produced. We begin by looking at the Descriptive Statistics table (see Figure 16.6), which presents the means and standard deviations for all groups. We can learn much about our whiteness study from the data presented in Figure 16.6.

Figure 16.6 Descriptive Statistics for the 2 × 2 ANOVA Whiteness Study

Descriptive Statistics

Dependent Variable:Whiteness meter reading

Type of detergent	Water temperature	Mean	Std. Deviation	N
Mighty Good	Hot	45.770	12.4716	10
	Cold	61.270	6.3659	10
	Total	53.520	[illegible]	20
Super max	Hot	52.240	[illegible]	[illegible]
	Cold	64.400	[illegible]	[illegible]
	Total	58.320	[illegible]	20
[illegible]	Hot	49.005	11.8698	20
	[illegible]	62.835	6.6528	20
	Total	55.920	11.8002	40

Total Whiteness level for hot water

Total whiteness level for cold water

One of the first things noticed in Figure 16.6 is the large difference in mean whiteness levels for hot and cold water (look at the bubbles in the table). These means of 49.005 (when clothes are washed in hot water) and 62.835 (for cold water) indicate that clothes washed in cold water are whiter. Further inspection of the table values shows that this is true for both the Mighty Good and Super Max detergents. The highest whiteness scores, 61.27 (Mighty Good) and 64.4 (Super Max), result from the cold water wash. Also look at the pairs of means for Mighty Good (45.77 and 61.27) and compare with Super Max (52.24 and 64.4) detergents. These whiteness values follow the same pattern relative to hot and cold water, so once again one might suspect detergent is not a principal factor in high readings on the whiteness meter. It could be said that we gained considerable insight from our descriptive analysis.

The next table selected from the Output Viewer is shown in Figure 16.7. The physical appearance of the table is small, but its importance is large when conducting our factorial analyses. When reading this table, you are interested in whether the null hypothesis is rejected. If the null hypothesis is rejected, then any significant differences identified by factorial analysis would be suspect, and it would be a good idea to use the nonparametric analog.

The null hypothesis for Levene's test is one of equal variances for all groups. Therefore, Levene's test is one of those cases where rejecting the null hypothesis would indicate that further analysis using two-way ANOVA is inappropriate. Two-way ANOVA requires equal variances for all groups. Figure 16.7 presents a significance level of .347, informing us that by failing to reject the null hypothesis, we have statistical evidence that the variances are equal across all groups. We may proceed with an increased level of confidence in any findings of our factorial analysis.

Figure 16.7 Levene's Test of Equality for Variances

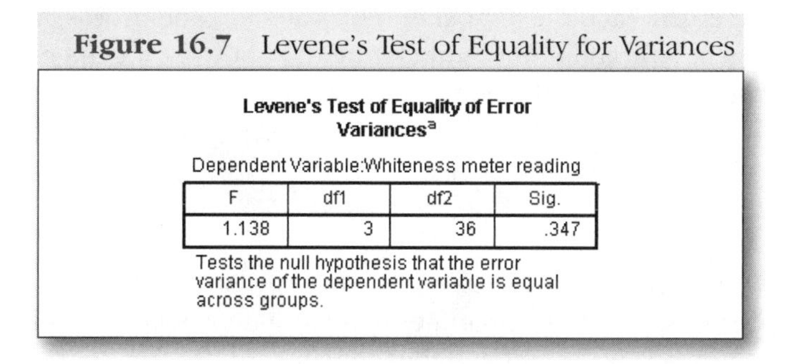

Levene's Test of Equality of Error Variances[a]

Dependent Variable:Whiteness meter reading

F	df1	df2	Sig.
1.138	3	36	.347

Tests the null hypothesis that the error variance of the dependent variable is equal across groups.

Figure 16.8 presents the most relevant information regarding whether the researcher can reject or fail to reject the null hypotheses. Differences in whiteness levels when using Mighty Good and Super Max detergents were found not to be significant, the level being .119. Since .119 is larger than .05, it can be said the differences are due to chance. The same nonsignificant outcome was identified when the interaction effects between detergents and water temperatures (detergent * water) were compared. See the "Sig." of .582 in Figure 16.8. Once again any interaction differences could only be attributed to chance. The significance level of .000 for the water temperature indicates that water temperature had a significant impact on the level of whiteness in the study.

Figure 16.8 ANOVA Table for Main and Interaction Effects in the Whiteness Study

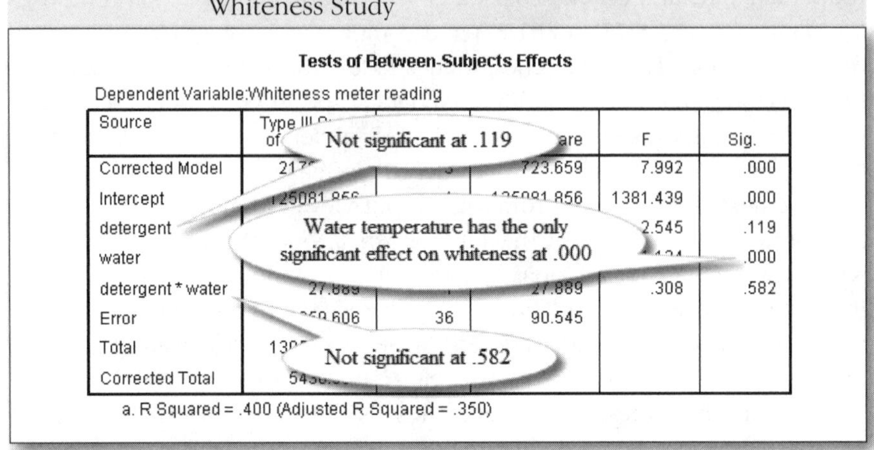

Tests of Between-Subjects Effects

Dependent Variable:Whiteness meter reading

Source	Type III [illegible] of [illegible]	[illegible]	[illegible]are	F	Sig.
Corrected Model	[illegible]	[illegible]	723.659	7.992	.000
Intercept	[illegible]	[illegible]	[illegible]	1381.439	.000
detergent	[illegible]	[illegible]	[illegible]	2.545	.119
water	[illegible]	[illegible]	[illegible]	[illegible]	.000
detergent * water	[illegible]	[illegible]	[illegible]	.308	.582
Error	[illegible]	36	90.545		
Total	[illegible]	[illegible]			
Corrected Total	[illegible]	[illegible]			

a. R Squared = .400 (Adjusted R Squared = .350)

What did we learn from our analysis of the data resulting from the clothes washing experiment? What did the two-way (factorial) ANOVA reveal? We can state that when clothes are washed in Mighty Good or Super Max, the level of whiteness does not significantly change. We can also state that when clothes are washed in cold water, the level of whiteness increases (see Figures 16.6 and 16.8). No interaction effect was identified for these two types of detergents and the temperature of water. In other words, we may say that detergent type and water temperature did not work together to change the level of whiteness by the end of the wash cycle.

△ 16.5 SUMMARY

This chapter expanded the concept of one-way ANOVA by permitting the investigator the latitude to examine multiple independent variables. When using two-way ANOVA, multiple independent variables can consist of many indivivdual groups. Such study designs can be extremely complex. An example that used two independent variables, each consisting of two groups, was used to demonstrate how SPSS analyzes such data. Chapter 17 introduces the reader to another form of the AVOVA test—repeated measures. The repeated measures procedure is used when the same subject or object is measured several times, usually over a period of time.

Chapter 17

One-Way ANOVA Repeated Measures Test and Friedman Test

17.1 Introduction and Objectives △

An AVOVA with repeated measures is a parametric test used for comparing three or more groups where the subjects are the same in each group. The ANOVA repeated measures test is often referred to as a within-subjects ANOVA. In a one-way ANOVA we are dealing with different independent groups, whereas in a repeated measures ANOVA we are dealing with the same group having repeated measures. Whereas in a paired-samples *t* test two levels are involved, in a repeated measures ANOVA, three or more levels are involved.

The Friedman test is a nonparametric test used as an alternative to the ANOVA repeated measures test when assumptions for the parametric test cannot be assumed. It is used to compare the means of the rankings of three or more matched groups.

OBJECTIVES

After completing this chapter, you will be able to

- Describe the purpose of the ANOVA repeated measures test
- Use SPSS to conduct an ANOVA repeated measures test
- Describe the assumptions related to an ANOVA repeated measures test
- Describe the purpose of the Friedman test
- Use SPSS to conduct the Friedman test
- Describe the assumptions related to the Friedman test

△ 17.2 RESEARCH SCENARIO AND TEST SELECTION

A statistics instructor wished to test his conjecture that the test scores of the 15 students in his class improved over time during the semester. The students' exam scores on three tests are listed in the table in Figure 17.1. The paired-samples *t* test is not appropriate because there are more than two groups. Consequently, the instructor selects the one-way ANOVA repeated measures test. We must use a repeated measures test because we are not dealing with three different groups of students. In contrast, we are dealing with the same group of students who have repeated measures, namely, three exams over time during the semester.

Figure 17.1 Student Scores on Three Exams Over Time

student	examone	examtwo	examthree
1	49	59	62
2	79	88	86
3	69	76	80
4	58	67	72
5	63	78	81
6	72	85	84
7	70	77	87
8	48	58	60

student	examone	examtwo	examthree
9	77	85	93
10	60	72	74
11	69	76	81
12	62	74	78
13	54	68	71
14	71	82	86
15	61	74	81

17.3 RESEARCH QUESTION AND NULL HYPOTHESIS

It may be expected that the students' scores on the three exams would improve over time as they become more acclimated to the subject matter and the instructor's approach to teaching. In contrast, it may be that scores might decrease as students become less interested in the subject or become overwhelmed by the increased complexity of the subject matter. The instructor assumes the former and he or she chooses the .05 level of significance. Assumptions of normal distributions and equivalent variances apply to the repeated measures test. Also, sphericity is assumed. Sphericity refers to the equality of the variances of the differences between levels of the repeated measures factor, in our case the three exams.

H_0: There is no significant difference in the mean test scores ($\mu_1 = \mu_2 = \mu_3$).

H_A: The mean scores are significantly different ($\mu_1 \neq \mu_2 \neq \mu_3$).

17.4 DATA INPUT, ANALYSIS, AND INTERPRETATION OF OUTPUT

You must enter the data in Figure 17.1 into SPSS before conducting the repeated measures test.

- Start SPSS.
- In the Variable View screen, enter the following three variables: **examone**, **examtwo**, and **examthree**.
- Set type to **numeric**.
- Set measure for each variable to **scale**.

- Set column to **8** and decimals to **0**.
- In the Data View screen, enter the data for each variable as shown in Figure 17.1.

Be certain to save this file as Repeat.

Now that you have entered variables and data, you can proceed to the analysis.

- Click **Analyze**, select **General Linear Model**, and click **Repeated Measures**, and a window titled Repeated Measures Define Factor(s) will open.
- Type **time** to replace **factor 1** in the Within Factor Subject Names: box.
- Enter **3** in the Number of Levels box and click **Add**.
- Click **Define**, and a window titled Repeated Measures will open.
- Click **examone**, and then click the right arrow to place it in the Within-Subjects Variables box. Do the same for **examtwo** and **examthree**.

If you click Post Hoc at this point to determine if there are significant differences in the exam scores, a window will open offering no options available, indicating that the options available for independent designs are not available for repeated measures designs. So we will take another approach.

- Click **options**, and a window titled Repeated Measures Options will open.
- Click **time** in the Factorial(s) and Factor Interaction box, and click the arrow to move it to the Display Means for box.
- Click **Compare main effects**.
- Select **LSD (none)** in the Confidence interval adjustment box.
- Click **Descriptive statistics**.
- Click **Continue**.
- Click **OK**.

Figure 17.2 shows the means and standard deviations for each exam.

Figure 17.2 Means and Standard Deviations for Three Exams

Descriptive Statistics

	Mean	Std. Deviation	N
examone	64.1333	9.41023	15
examtwo	74.6000	8.86244	15
examthree	78.4000	9.17917	15

Figure 17.3 shows the results of Mauchly's test for equality of variances. The level of significance, .680, is greater than .05, indicating that the null hypothesis of nonequivalence of variances is rejected. Sphericity may be assumed. Consequently, we can disregard the Greenhouse-Geisser and Huynh-Feldt tests, which are corrections for violations of sphericity.

Figure 17.3 Mauchly's Test of Sphericity

Mauchly's Test of Sphericity[b]

Measure:MEASURE_1

Within Subjects Effect	Mauchly's W	Approx. Chi-Square	df	Sig.	Epsilon[a]		
					Greenhouse-Geisser	Huynh-Feldt	Lower-bound
time	.942	.770	2	.680	.946	1.000	.500

Tests the null hypothesis that the error covariance matrix of the orthonormalized transformed dependent variables is proportional to an identity matrix.

Figure 17.4 shows the results of the tests of within-subjects effects. We are concerned with the row indicating time and sphericity assumed as this gives the level of significance, which indicates whether the means of the three exams are significantly different. The *F* value is 180, and the level of significance is .000. Since .000 is less than .05, we reject the null hypothesis that the means are not significantly different. That is, the exam scores of the students over time were significantly different.

Figure 17.4 Tests of Within-Subjects Effects

Tests of Within-Subjects Effects

Measure:MEASURE_1

Source		Type III Sum of Squares	df	Mean Square	F	Sig.
time	Sphericity Assumed	1637.644	2	818.822	180.496	.000
	Greenhouse-Geisser	1637.644	1.891	865.930	180.496	.000
	Huynh-Feldt	1637.644	2.000	818.822	180.496	.000
	Lower-bound	1637.644	1.000	1637.644	180.496	.000
Error(time)	Sphericity Assumed	127.022	28	4.537		
	Greenhouse-Geisser	127.022	26.477	4.797		
	Huynh-Feldt	127.022	28.000	4.537		
	Lower-bound	127.022	14.000	9.073		

Figure 17.5 shows the results of the pairwise comparisons of the three exams. Note that some comparisons are repetitive. Since .000 is less than .05, the mean difference score for each pair of exams is significantly different. Consequently, we have statistical evidence that the mean scores shown in Figure 17.2 are significantly different. The mean of Exam 1 is greater than the mean of Exam 2, and the mean of Exam 2 is greater than the mean of Exam 3.

Figure 17.5 Pairwise Comparisons of the Three Exams

Pairwise Comparisons

Measure:MEASURE_1

(I) time	(J) time	Mean Difference (I-J)	Std. Error	Sig.[a]	95% Confidence Interval for Difference[a]	
					Lower Bound	Upper Bound
1	2	-10.467*	.682	.000	-11.930	-9.003
	3	-14.267*	.842	.000	-16.073	-12.460
2	1	10.467*	.682	.000	9.003	11.930
	3	-3.800*	.800	.000	-5.516	-2.084
3	1	14.267*	.842	.000	12.460	16.073
	2	3.800*	.800	.000	2.084	5.516

Based on estimated marginal means

*. The mean difference is significant at the .05 level.

△ 17.5 NONPARAMETRIC TEST: FRIEDMAN TEST

The data listed in Figure 17.1 can be shown to meet all the assumptions required for a parametric test (scale data, normally distributed, and equal variances). However, we wish to demonstrate the Friedman test available in SPSS when these assumptions are not met. Since the Friedman test is based on ranking of data, we show an extended version of the table displayed in Figure 17.1 in which we include the rankings for each student for the three exams (1 = *lower*, 2 = *middle*, and 3 = *higher*). See Figure 17.6. The mean ranks are computed by summing the ranks for each column and dividing by 15.

Figure 17.6 Means and Ranks for the Three Exams

student	examone	examtwo	examthree	Rank	Rank	Rank
1	49	59	62	1	2	3
2	79	88	86	1	3	2
3	69	76	80	1	2	3

student	examone	examtwo	examthree	Rank	Rank	Rank
4	58	67	72	1	2	3
5	63	78	81	1	2	3
6	72	85	84	1	3	2
7	70	77	87	1	2	3
8	48	58	60	1	2	3
9	77	85	93	1	2	3
10	60	72	74	1	2	3
11	69	76	81	1	2	3
12	62	74	78	1	2	3
13	54	68	71	1	2	3
14	71	82	86	1	2	3
15	61	74	81	1	2	3
	Mean = 64.13	Mean = 74.60	Mean = 78.40	Mean Rank = 1.00	Mean Rank = 2.13	Mean Rank = 2.87

Follow these steps to perform the Friedman analysis:

- Start SPSS.
- Click **Analyze**, click **Nonparametric Tests**, and then click **Related Samples**. A window titled "Non Parametric Tests: Two or More Related Samples" will open.
- Click **Automatically compare observed data to hypothesized**.
- Click **Fields** and then click **Use Custom Field Assignments**.
- Click **examone** in the Fields panel, and click the right arrow to place it in the Test Fields panel. Do the same for examtwo and examthree.
- Click **Settings**, and then click **Customize Tests**.
- Click **Friedman's two-way ANOVA by ranks (k samples)**.
- Choose **All pairwise** from the Multiple comparisons box and click **Run**.
- A table titled Hypothesis Test Summary will appear in the Output Viewer.
- Double click on this table, and a window will open consisting of two panels, showing the expanded Model Viewer and the Model Viewer on which you just double clicked.
- At the bottom of the expanded Model Viewer, click **View** and then select **Pairwise Comparisons**.

The tables displayed in the Output Viewer are shown in Figures 17.7, 17.8, and 17.9.

The null hypothesis in the Friedman test is that distributions of the ranks of the exam scores are the same. In Figure 17.7, since .000 is less than .05, we reject the null hypothesis of no difference indicating that there is a significant difference in the mean ranks of the three exams. This result agrees with that of the earlier parametric test.

Figure 17.8 provides more informative statistics, including the mean ranks as follows: examone, 1.00; examtwo, 2.13; and examthree, 2.87. Inspection of the results shown in Figure 17.8 gives a test statistic of 26.53 and a significance level of .000. Since .000 is less than .05, we reject the null hypothesis that there is no difference in the rank means of the three exams, as we did on inspection of the results shown in Figure 17.7.

Figure 17.7 Hypothesis Test Summary for Friedman Test

Hypothesis Test Summary

	Null Hypothesis	Test	Sig.	Decision
1	The distributions of examone, examtwo and examthree are the same.	Related-Samples Friedman's Two-Way Analysis of Variance by Ranks	.000	Reject the null hypothesis.

Asymptotic significances are displayed. The significance level is .05.

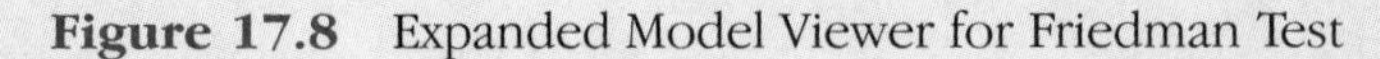

Figure 17.8 Expanded Model Viewer for Friedman Test

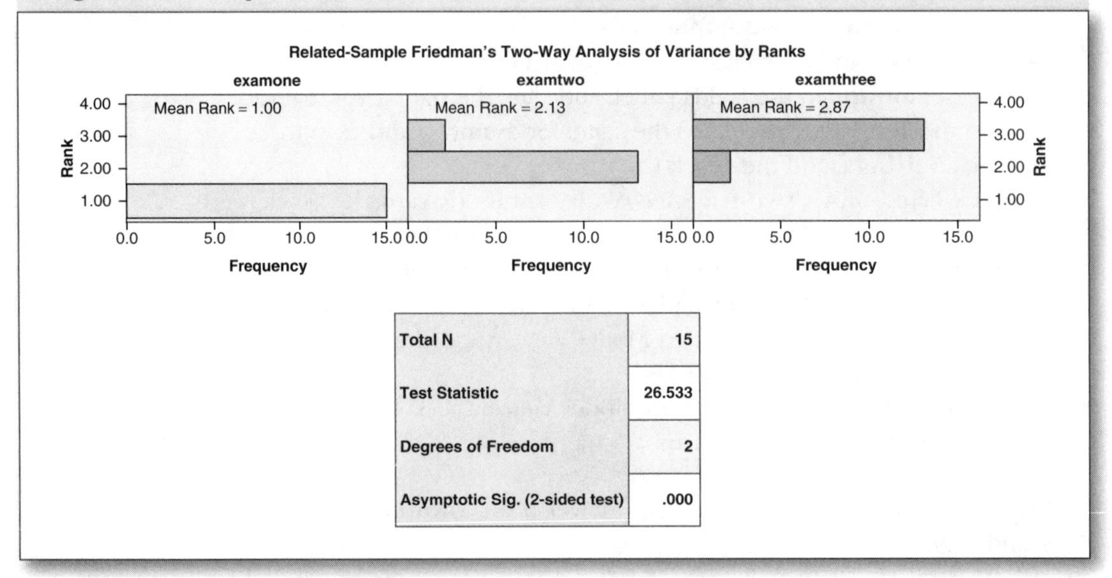

Total N	15
Test Statistic	26.533
Degrees of Freedom	2
Asymptotic Sig. (2-sided test)	.000

Figure 17.9 displays the results regarding the comparisons of the mean ranks for the three exams. Since .006 is less than .05 we conclude that the mean rank of Exam 2 is significantly higher than the mean rank of Exam 1. However, in contrast to the results of the parametric test of means, since .134 is greater than .05, we conclude that the mean rank of Exam 2 is not greater than the mean rank of Exam 3. The finding of no difference for this one comparison illustrates that that the power of a nonparametric test to detect differences is less.

Figure 17.9 Pairwise Comparisons for Friedman Test

Sample1-Sample2	Test Statistic	Std. Error	Std. Test Statistic	Sig.	Adj.Sig.
examone-examtwo	-1.133	.365	-3.104	.002	.006
examone-examthree	-1.867	.365	-5.112	.000	.000
examtwo-examthree	-.733	.365	-2.008	.045	.134

Each row tests the null hypothesis that the Sample 1 and Sample 2 distributions are the same.
Asymptotic significances (2-sided tests) are displayed. The significance level is .05.

17.6 SUMMARY

The one-way ANOVA repeated measures test is appropriate when comparing three or more groups when the subjects undergo repeated measures. It is equivalent to a paired-samples *t* test where only two groups are involved. If assumptions for the ANOVA cannot be assumed, then the nonparametric Friedman test can be used. However, it should be remembered that the power of a nonparametric test is less than that of a parametric test.

CHAPTER 18

ANALYSIS OF COVARIANCE (ANCOVA)

△ 18.1 INTRODUCTION AND OBJECTIVES

The analysis of covariance (ANCOVA) statistical test is a method of statistically equating groups on one or more variables and for increasing the power of a statistical test. Basically, ANCOVA allows the researcher to remove the effect of a known concomitant or control variable know as a covariate. We include a covariate in the analysis to account for the effect of a variable that does affect the dependent variable but that could not be accounted for in the experimental design. In other words, we use a covariate to adjust scores on a dependent variable for initial differences in another variable. For example, in a pretest, posttest analysis, we can use the pretest as a covariate to statistically control for the pretest scores, meaning that we statistically equate all participants on their pretest scores after which we examine their posttest scores. ANCOVA essentially tests whether certain factors have an effect on the dependent variable after removing the variance accounted for by the covariate.

There are some rather strict assumptions regarding the use of ANCOVA. All assumptions regarding the analysis of variance (ANOVA) are required. In addition, the covariate should have a reasonable correlation with the dependent variable, meaning that there is a linear relationship between the covariate and the independent variable. The ANCOVA must also satisfy the additional assumption of homogeneous regression slopes, meaning that the slopes of the regression lines representing the relationship between covariate and dependent variable are similar. See Appendix D for further explanation.

We realize that we have presented some rather difficult concepts regarding ANCOVA, but these are necessary for a variety of reasons, one being that ANCOVA is often used inappropriately because assumptions have not been met, especially the assumption of homogeneity of regression slopes. With that said, there are occasions when ANCOVA seems to offer the only viable method of control, albeit statistical, open to the researcher.

OBJECTIVES

After completing this chapter, you will be able to

- Describe the assumptions required for use of the analysis of covariance
- Describe the purpose of a covariate in the analysis of covariance
- Use SPSS to perform an analysis of covariance
- Interpret the results of an analysis of covariance

18.2 Research Scenario and Test Selection

A researcher wished to make a preliminary determination regarding which of four novel methods of teaching reading available to a school district might be more effective for third-grade students. He randomly selected 24 third-grade students and then randomly assigned 6 to each of the 4 methods of reading instruction. A pretest in reading was administered to the 24 students. After several months of reading instruction, a posttest was administered. Figure 18.1 displays the results.

Figure 18.1 Data for Four Methods of Teaching Reading

Reading Method One		Reading Method Two		Reading Method Three		Reading Method Four	
Pretest	Posttest	Pretest	Posttest	Pretest	Posttest	Pretest	Posttest
25	160	27	175	19	151	36	196
26	165	27	164	27	184	27	168
18	125	22	166	20	133	32	195
20	151	21	157	30	185	31	188
28	162	27	175	21	155	19	137
24	146	22	165	24	167	31	184

△ 18.3 RESEARCH QUESTION AND NULL HYPOTHESIS

Since differences in pretest scores might contribute to differences in posttest scores, the researcher decided to use pretest scores as a covariate. Consequently, an analysis of covariance is the appropriate statistical test.

We state the null and alternative hypotheses as follows:

H_0: There is no significant difference in the four methods of teaching reading ($\mu_1 = \mu_2 = \mu_3 = \mu_4$).

H_A: There is a significant difference in the four methods of teaching reading ($\mu_1 \neq \mu_2 \neq \mu_3 \neq \mu_4$).

△ 18.4 DATA INPUT, ANALYSIS, AND INTERPRETATION OF OUTPUT

You will code Reading Method One as 1, Reading Method Two as 2, Reading Method Three as 3, and Reading Method Four as 4. There are three variables involved: Reading Method, Pretest, and Posttest.

- Start SPSS.
- Click **File**, Select **New**, and click **Data**.
- Click the **Variable View** screen.
- In the first cell below *Name* enter **ReadMethod.**
- In the first cell below *Type*, select **Numeric** and set decimals to **0**.
- In the first cell below *Decimals*, select **0**.
- In the first cell below *Label*, enter **Reading Method**.
- Click the first cell below *Values*, and a window titled Value Labels will open.
- Enter **1** for value and **Reading Method One** for label, and click **Add**.
- Repeat as follows: **2** and **Reading Method Two**, **3** and **Reading Method Three**, **4** and **Reading Method Four**. Click **OK**.
- In the second cell below *Name*, enter **Pretest**.
- In the second cell below *Type*, enter **Numeric** and set decimals to **0**.
- In the second cell below *Decimals*, select **0**.
- In the second cell below *Label*, enter **Pretest Points**.
- In the second cell below *Measure*, select **Scale**.
- In the third cell below *Name*, enter **Posttest**.
- In the third cell below *Type*, enter **Numeric** and set decimals to **0**.
- In the third cell below *Decimals*, select **0**.

- In the third cell below *Label*, enter **Posttest Points**.
- In the third cell below *Measure*, select **Scale**.

Now that you have entered the variables and values, enter the data from Figure 18.1.

- Click the **Data View** screen.
- In the first cell below ReadMethod, enter **1**.
- In the first cell below Pretest, enter **25**.
- In the first cell below Posttest, enter **160**.
- In the second cell below ReadMethod, enter **1**.
- In the second cell below Pretest, enter **26**.
- In the second cell below Posttest, enter **165**.
- In the third cell below ReadMethod, enter **1**.
- In the third cell below Pretest, enter **18**.
- In the third cell below Posttest, enter **125**.
- In the fourth cell below ReadMethod, enter **1**.
- In the fourth cell below Pretest, enter **20**.
- In the fourth cell below Posttest, enter **151**.
- In the fifth cell below ReadMethod, enter **1**.
- In the fifth cell below Pretest, enter **28**.
- In the fifth cell below Posttest, enter **162**.
- In the sixth cell below ReadMethod, enter **1**.
- In the sixth cell below Pretest, enter **24**.
- In the sixth cell below Posttest, enter **146**.
- In the seventh cell below ReadMethod, enter **2**.
- In the seventh cell below Pretest, enter **27**.
- In the seventh cell below Posttest, enter **175**.

In the same manner, continue entering data for Reading Methods Two, Three, and Four. After you have completed all entries, be certain to save this file now as **covar**. You will need this file if you wish to perform the SPSS analysis for homogeneity of regression slopes as describe in Appendix D.

After you have entered the variables, values, and data, do the following:

- Start SPSS.
- Click **Analyze**, select **General Linear Model**, and then click **Univariate**. A window titled Univariate will open.
- Click **Posttest Points** and click the right arrow to move it to the Dependent Variable box.
- Click **Reading Method** and click the right arrow to move it to the Fixed Factor(s) box.
- Click **Pretest Points** and click the right arrow to move it to the Covariate(s) box.

- Click **Options**, click **Descriptive Statistics**, and then click **Homogeneity Tests**.
- Click **Continue**, and then click **OK**.

Figure 18.2 displays the mean scores for the four methods of teaching reading.

Figure 18.2 Means for the Four Methods of Teaching Reading

Descriptive Statistics

Dependent Variable:Posttest Points

Reading Method	Mean	Std. Deviation	N
Reading Method One	151.50	14.816	6
Reading Method Two	167.00	6.957	6
Reading Method Three	162.50	20.236	6
Reading Method Four	178.00	22.494	6
Total	164.75	18.748	24

Levene's test shown in Figure 18.3 indicates that the assumption of equality of variances has not been violated, because .728 is greater than .05, meaning that we fail to reject the null hypothesis, which states that the variances are equal.

Figure 18.3 Levene's Test for Equality of Variances

Dependent Variable:Posttest Points

F	df1	df2	Sig.
.438	3	20	.728

Tests the null hypothesis that the error variance of the dependent variable is equal across groups.

Note: SPSS provides a test that requires a bit of customizing to determine if regression slopes are homogeneous. We applied this test, and it indicated that regression slopes are, indeed, homogeneous. See Appendix D, if you wish to run the test for homogeneity.

Figure 18.4 gives *F* values, degrees of freedom, and significance levels for all main effects and the covariate. We wish to determine from the results listed in the table if there is a difference in the four methods of teaching reading after the pretest is accounted for (after we have statistically removed the effect of the pretest).

Let's inspect the data in the table shown in Figure 18.4. Since .000 is less than .05, we conclude that the researcher's idea that differences in pretest scores might contribute to differences was correct. We reject the null hypothesis that states pretest scores are not significantly related to posttest scores.

The main effect for Reading Method is significant (.040 is less than .05), indicating that there are statistically significant differences in the four methods of teaching reading after adjusting for the effect of the pretest.

Figure 18.4 Results of Covariance Analysis

Tests of Between-Subjects Effects

Dependent Variable:Posttest Points

Source	Type III Sum of Squares	df	Mean Square	F	Sig.
Corrected Model	6993.184[a]	4	1748.296	30.438	.000
Intercept	3254.488	1	3254.488	56.661	.000
Pretest	4825.684	1	4825.684	84.016	.000
ReadMethod	579.914	3	193.305	3.365	.040
Error	1091.316	19	57.438		
Total	659506.000	24			
Corrected Total	8084.500	23			

a. R Squared = .865 (Adjusted R Squared = .837)

You can perform an ANOVA on the posttest scores alone as shown in Figure 18.5. Interestingly, the results of the ANOVA indicate that there is no significant difference among the posttest scores (.094 is greater than .05), meaning we fail to reject the null hypothesis. The methods of teaching reading are equally effective. This result is in direct contrast to the ANCOVA results, which indicated that the four methods of teaching reading are not equally effective. Including the pretest as a covariate increased the power of this test.

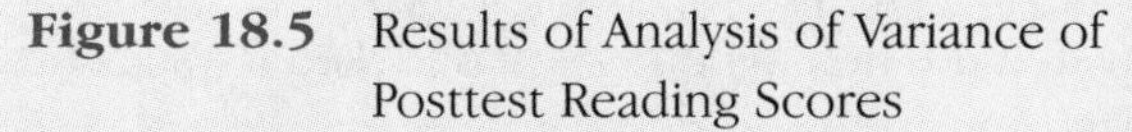

Figure 18.5 Results of Analysis of Variance of Posttest Reading Scores

ANOVA

Posttest Points

	Sum of Squares	df	Mean Square	F	Sig.
Between Groups	2167.500	3	722.500	2.442	.094
Within Groups	5917.000	20	295.850		
Total	8084.500	23			

18.5 SUMMARY

The ANCOVA is used to statistically control for factors that cannot be, or were not, controlled for methodologically by methods such as random assignment. If the researcher has evidence that a factor or factors cannot be controlled for methodologically but might have an effect on the dependent variable, he or she should statistically control for these factors via an ANCOVA. Such factors should have a linear relationship to the independent variable.

Chapter 19

Pearson's Correlation and Spearman's Correlation

19.1 Introduction and Objectives

The preceding chapters on hypothesis testing have presented methods intended to give SPSS users tools that will assist them in fulfilling one of the major goals in statistics—the establishment of relationships between variables. That goal was pursued by comparing group means, variances, and medians. We tested the null hypothesis and looked for statistically significant differences between various groups.

There are many situations in statistics where you are not comparing the scores of groups and cannot calculate means and variances. An example might be that you are interested in studying the relationship between IQ and income level within a single sample of individuals. A situation where you are comparing two variables calls for the use of a descriptive statistic known as a correlation coefficient.

We define a correlation coefficient as a numerical value and descriptive statistic that indicates the degree (strength) to which two variables are related and whether any detected relationship is positive or negative (direction). A positive relationship results if one variable increases and the other does also. A negative relationship is such that as one variable increases, the other decreases.

Two important descriptive functions of both the parametric and nonparametric correlation coefficients are to determine the *strength* and *direction* of the relationship between two variables. To understand these two aspects of correlation, let's examine the hypothetical scale data presented in Figure 19.1.

Figure 19.1 Correlated Data for Two Variables

Name	Hours of Study	Points Earned
David	1	25
Julie	2	50
Mike	3	75
Maricela	4	100

The data in Figure 19.1 show a relationship because as one variable increases so does the other (hours of study increases and points earned also increases). We have evidence of a strong relationship since each 1 hour of study resulted in the student earning an additional 25 points. Furthermore, we say that the direction is positive since both variables increase together. The calculated correlation coefficient quantifies the strength and specifies the direction. If we used SPSS to calculate the correlation coefficient, we would find its value to be +1.00. What does this value mean?

The plus sign indicates a positive correlation, whereas the value of 1.00 indicates a perfect correlation between hours of study and points earned on the exam. A calculated correlation coefficient can take on any number between 0 and a positive or negative 1 (e.g., .43, .95, .88, −.35, −.99, or −.06). The closer the positive or negative correlation coefficient is to a +1.00 or −1.00, the stronger the relationship. A correlation coefficient of zero indicates that there is no relationship between the two variables.

There are many types of correlation coefficients that are designed for use with different levels of measurement. Two of the most commonly used correlation coefficients are Pearson's and Spearman's. These are the subject of our chapter on correlation. Let's take a brief look at both these correlation coefficients.

Pearson's product-moment correlation coefficient is used when both the variables are measured at the scale level (interval/ratio). The variables must also be approximately normally distributed and have a linear relationship. Thus, the Pearson's coefficient is referred to as the parametric descriptive statistic. The data presented in Figure 19.1 fulfill these requirements; therefore, the appropriate correlation coefficient would be Pearson's correlation coefficient.

When either variables are measured at the ordinal level or your scale data are not normally distributed, you should use the nonparametric Spearman's correlation coefficient. For example, a researcher is interested in studying a survey instrument consisting of ranked responses. One variable ranks the importance of religion as very important, important, neutral, of little importance, and not at all important. The second variable ranks educational level as high school degree, associate degree, bachelor's degree, master's degree, or doctoral degree. To investigate the strength of the relationship between religion and education, the nonparametric Spearman's correlation coefficient is the correct statistic.

It must be mentioned that no matter how strong the calculated correlation coefficient might be, you cannot infer a causal relationship. Just because two variables move together in a linear manner does not indicate that one causes the change in the other. One of the most often misused statistics is when a high correlation coefficient is used to draw the conclusion that one variable is the cause of an increase (or decrease) in the other. The data in Figure 19.1 are a prime example, in that just by observing the strong relationship we cannot say that increased study causes the increase in the number of points earned on the exam. From the information provided in the hours of study and points earned example, we have no information about the multitude of other variables that may have affected the points earned on the exams. We must not infer a causal relationship between the two variables.

Significance Test

SPSS automatically calculates a significance level for the correlation coefficient. SPSS assumes that the cases were selected at random from a larger population. The research question asks whether the calculated correlation coefficient can be taken seriously. When we say "taken seriously," we are simply asking whether the sample correlation coefficient is representative of the population's correlation coefficient value or the sample's value occurred by chance.

The alternative hypothesis states that the population correlation coefficient does *not* equal zero, and we write H_A: $\rho \neq 0$. In the expression just given, ρ (rho) is the hypothesized population correlation coefficient. The null hypothesis is that the population has a true correlation coefficient of zero, H_0: $\rho = 0$. If the null hypothesis can be rejected, then we have some evidence that the calculated correlation is not due to chance—we can take it seriously.

OBJECTIVES

After completing this chapter, you will be able to

Define and describe the correlation coefficient

Describe the data assumptions required for Pearson's correlation coefficient

Explain the purpose of computing correlation coefficients

Use SPSS to calculate Pearson's and Spearman's correlation coefficients

Describe and interpret results from Pearson's correlation coefficient

Use SPSS to conduct the test of significance for Pearson's correlation coefficient

Interpret the significance test for Pearson's correlation coefficient

Describe data assumptions required for Spearman's correlation coefficient

Describe and interpret results from Spearman's correlation coefficient

19.2 PEARSON'S PRODUCT-MOMENT CORRELATION COEFFICIENT △

We use the **class_survey1.sav** database to demonstrate the Pearson's correlation coefficient. In the class survey database, there are two variables that recorded the points earned on two exams. You will use SPSS to determine the strength and direction of the relationship between the student scores on these exams. Furthermore, we regard these exam scores as representative of a larger population. Therefore, we will interpret SPSS's test of significance in an effort to provide evidence that the sample correlation coefficient represents the unknown population correlation coefficient.

Check for Linear Relationship Between Variables

First, we check for a linear relationship between the two exams by using SPSS's Chart Builder. Then, we proceed with the test for normality for both variables.

- Start SPSS.
- Click **Cancel** in the SPSS Statistics opening window.

- Click **File**, select **New**, and click **Data**.
- Click **class_survey1.sav**, and then click **Open**.
- Click **Graphs**, and then click **Chart Builder** (the Chart Builder window opens; see Figure 19.2).
- Click **Scatter/Dot**, and then double click **Simple Scatter** (which moves this chart style to the Chart preview pane).
- Click and drag **Points on Exam One** to x-axis in the Chart preview pane.
- Click and drag **Points on Exam Two** to y-axis in the Chart Builder pane.
- Click **OK** (the Output Viewer opens; see Figure 19.3).

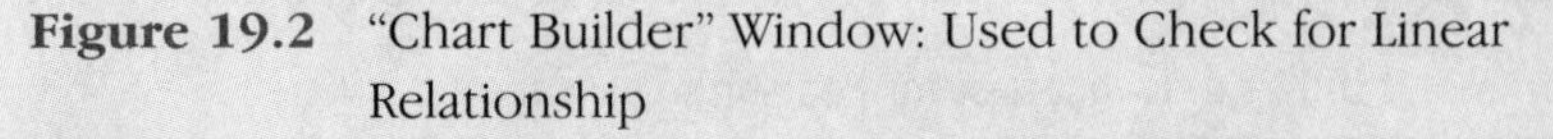

Figure 19.2 "Chart Builder" Window: Used to Check for Linear Relationship

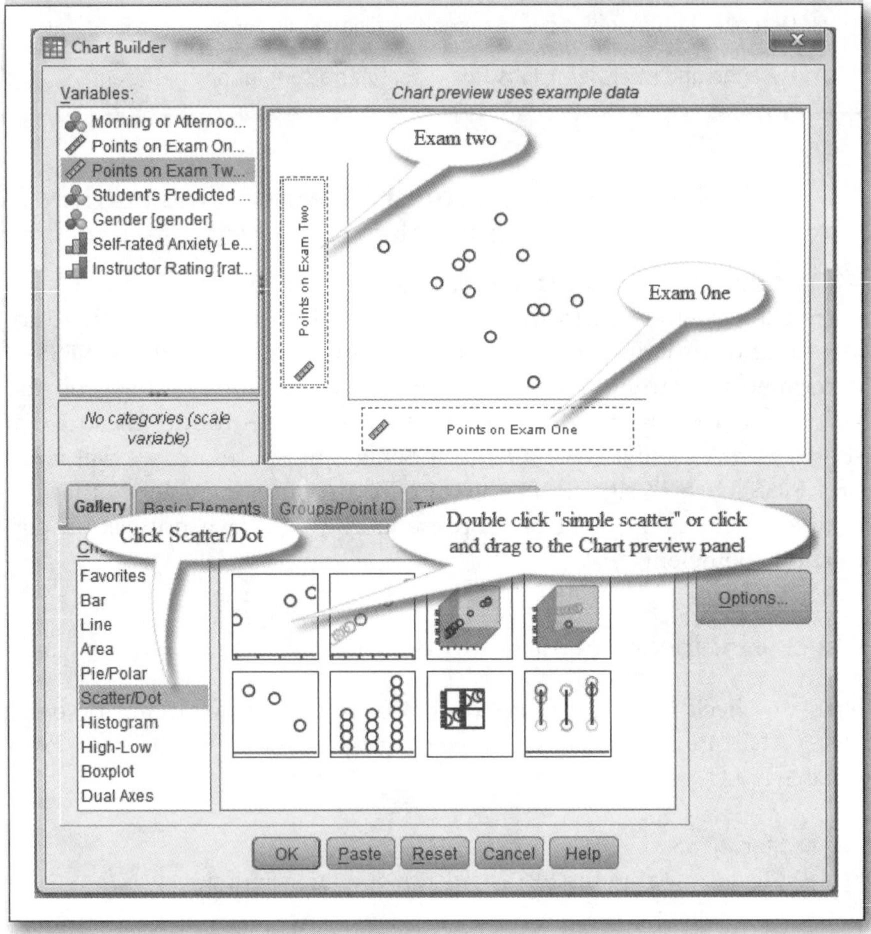

Figure 19.3 Scatter Plot Showing a Weak Linear Relationship Between Exams

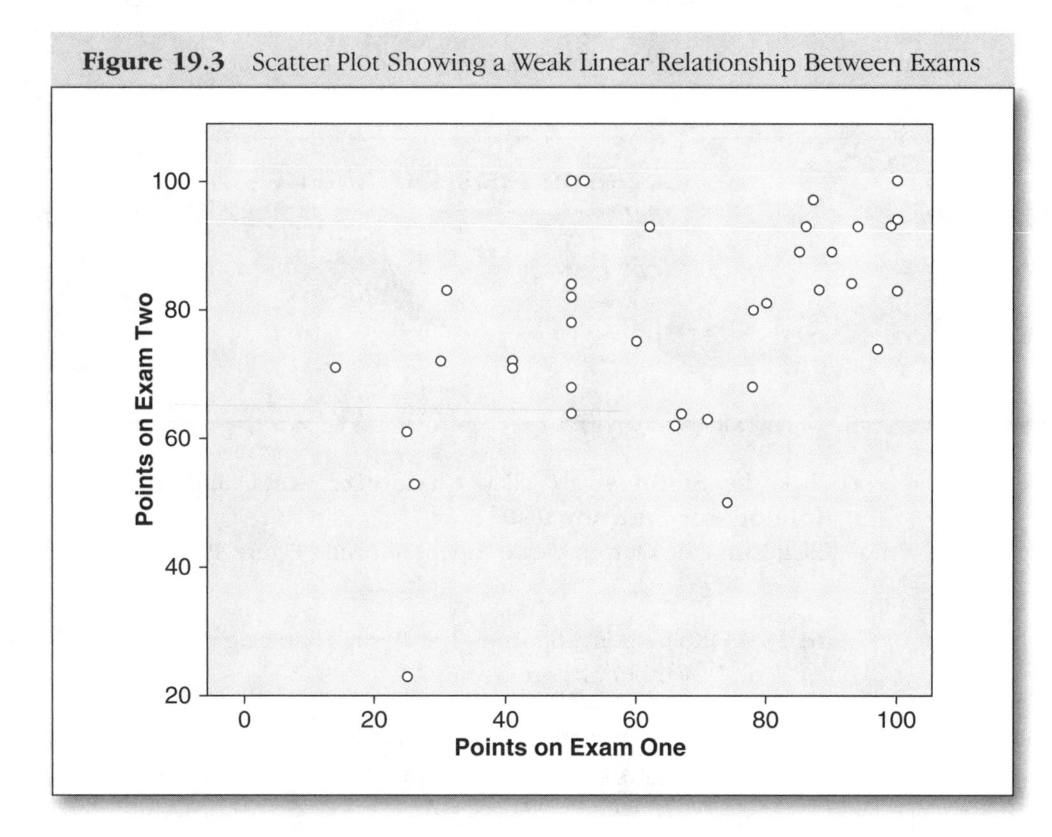

Figure 19.3 shows that there is a weak linear relationship between these two variables. We next check both variables to see if they are approximately normally distributed.

Check Both Variables for Normality

The checks for normality involve the use of the nonparametric Kolmogorov-Smirnov test as presented below.

- Click **Analyze**, select **Nonparametric Tests**, and then click **One-Sample** (the "One-Sample Nonparametric Tests" window opens—it has three tabs at its top: Objective, Fields, and Settings; see Figure 19.4).
- Click the **Objective** tab, and then click **Customize analysis**.
- Click the **Fields** tab (make sure that **Points on Exam One** and **Points on Exam Two** are the only variables in the Test Fields panel as shown in Figure 19.4). (Note: You may have to click variables and use the arrow to accomplish this task.)

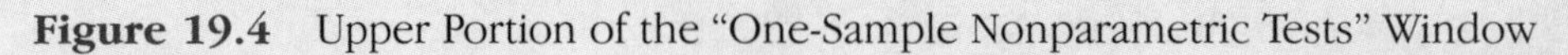

Figure 19.4 Upper Portion of the "One-Sample Nonparametric Tests" Window

- Click the **Settings** tab, click **Customize tests**, and then click **Kolmogorov-Smirnov test**.
- Click **Run** (the Output Viewer opens showing Figure 19.5).

Figure 19.5 Kolmogorov-Smirnov Test Results Showing Both Distributions Are Normal

Hypothesis Test Summary

	Null Hypothesis	Test	Sig.	Decision
1	The distribution of Points on Exam One is normal with mean 65.946 and standard deviation 25.951.	One-Sample Kolmogorov-Smirnov Test	.662	Retain the null hypothesis.
2	The distribution of Points on Exam Two is normal with mean 78.108 and standard deviation 16.551.	One-Sample Kolmogorov-Smirnov Test	.906	Retain the null hypothesis.

Asymptotic significances are displayed. The significance level is .05.

The results of the Kolmogorov-Smirnov test are satisfactory as both distributions are determined to be normal. The significance levels are .662 (exam 1) and .906 (exam 2); both values exceed .05, so therefore we fail to reject the null hypothesis that the distributions are normal. We may now calculate the correlation coefficient and the hypothesis test.

Calculation of the Pearson's Correlation Coefficient and Test of Significance

- Click **Analyze**, select **Correlate**, and then click **Bivariate** (the "Bivariate Correlations" window opens; see Figure 19.6).

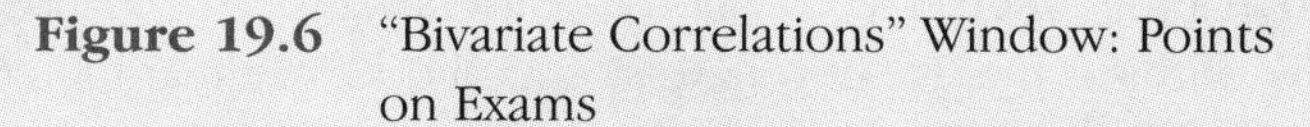

Figure 19.6 "Bivariate Correlations" Window: Points on Exams

- Click **Points on Exam One**, and then click the arrow.
- Click **Points on Exam Two**, and then click the arrow.
- Click **Pearson** (if not already checked).
- Click **Flag significant correlations** (if not already checked).
- Click **OK** (the Output Viewer opens; see Figure 19.7).

Figure 19.7 Pearson's Correlation Output: Exams One and Two

Correlations

		Points on Exam One	Points on Exam Two
Points on Exam One	Pearson Correlation	1	.536**
	Sig. (2-tailed)		.001
	N	37	37
Points on Exam Two	Pearson Correlation	.536**	1
	Sig. (2-tailed)	.001	
	N	37	37

**. Correlation is significant at the 0.01 level (2-tailed).

Figure 19.7 shows a correlation of .536 when Exam One and Exam Two are compared. We can say they are moderately correlated in a positive direction. Figure 19.7 also indicates that the correlation coefficient of .536 is significant at the .001 level. The interpretation is that there is a very small probability

(1 in 1,000) that the observed correlation coefficient was due to chance. Another way to think of this is that we now have statistical evidence that the population correlation coefficient does not equal zero. We reject the null hypothesis H_0: $\rho = 0$ (remembering that ρ = the hypothesized population correlation coefficient), and we now have evidence in support of the alternative hypothesis, H_A: $\rho \neq 0$. We can take the correlation coefficient of .536 seriously (meaning that it did not occur by chance). We next address the use of Spearman's correlation coefficient.

△ 19.3 NONPARAMETRIC TEST: SPEARMAN'S CORRELATION COEFFICIENT

Spearman's correlation coefficient is a statistic that shows the strength of the relationship between two variables measured at the ordinal level. The Spearman's correlation coefficient can also be used as the nonparametric alternative to the Pearson's correlation coefficient if scale data are not normally distributed.

If the **class_survey1.sav** database is not open, locate and open it because it is used to demonstrate Spearman's correlation coefficient as applied to two ordinal variables.

- Start SPSS.
- Click **Cancel** in the SPSS Statistics opening window.
- Click **File**, select **New**, and click **Data**.
- Click **class_survey1.sav**, and then click **Open**.
- Click **Analyze**, select **Correlate**, and then click **Bivariate** (the "Bivariate Correlations" window opens; see Figure 19.6 in the previous parametric section).
- Click **Self-rated Anxiety**, and then click the arrow.
- Click **Instructor Rating**, and then click the arrow.
- Unclick **Pearson**, and then click **Spearman**.
- Click **OK** (the Output Viewer opens; see Figure 19.8).

Figure 19.8 Spearman Correlation Output: Self-Rated Anxiety and Instructor Rating

Correlations

			Self-rated Anxiety Level	Instructor Rating
Spearman's rho	Self-rated Anxiety Level	Correlation Coefficient	1.000	-.073
		Sig. (2-tailed)	.	.669
		N	37	37
	Instructor Rating	Correlation Coefficient	[illegible]	1.000
		Sig. (2-tailed)	[illegible]	.
		N	37	37

Very low

Not significant

Instructor Rating and Self-rated Anxiety Level are the two variables chosen to demonstrate Spearman's correlation coefficient. We are interested in studying these data to determine if there was a relationship between how a student rated the instructor's performance and his or her self-rated level of anxiety.

Figure 19.8 reports that the Spearman's correlation coefficient is −.073 with a significance level of .669. What does this mean for our investigation of Instructor Rating and Self-rated Anxiety Level? First of all, the r_s of −.073 informs us that there is no relationship between these two ordinal variables. Knowing how a person rated the instructor would be of no help in predicting his or her level of anxiety or vice versa. We also note that the null hypothesis, which stated that the population correlation coefficient equalled zero, was not rejected. From this, we conclude that our finding of a low correlation coefficient could be attributed to chance. We are unable to take the calculated correlation seriously, and additional study would be advised.

19.4 SUMMARY △

This chapter placed an emphasis on the analysis of data already collected. We studied the relationships between two variables and then answered questions regarding the strength and direction of any detected relationship. As a statistical technique, correlation is descriptive rather than inferential. However, there was an inferential aspect presented in our correlation chapter. Hypothesis tests were used to answer the question whether the calculated correlation coefficient is representative of what exists in the population. Stated succinctly, the question is, can we take the observed sample correlation seriously or could it be the result of chance? In the following chapter, we extend these correlation concepts to regression analysis.

Chapter 20

Single Linear Regression

△ 20.1 Introduction and Objectives

Linear regression is actually an extension of Pearson's (linear) correlation. In Chapter 19, we demonstrated how two variables could be described in terms of the strength and direction of their relationship. We also discussed how linear correlation could assist us in predicting the value of one variable through the knowledge of the other. An example of such prediction would be that the manager of a public swimming pool could observe a day's temperature and record the number of patrons at the pool. Several days could be observed, and the information could be used to estimate the number of expected patrons for any day's temperature just by looking at data in an informal way. Let's explain how this informal use of correlation might be expanded to single linear regression—the subject of this chapter.

Single linear regression is a statistical technique that describes the relationship between two variables by the calculation of a prediction equation. The prediction equation averages all prior observed relationships between two variables. These averages are then used to develop a precise equation to predict unknown values of the dependent variable.

The descriptor "single" in the linear regression name refers to the number of independent variables. In single linear regression, there is one independent variable used to predict the value of the dependent variable. Single variable regression is contrasted with multiple regression, where you have two or more independent variables. Multiple regression is discussed in Chapter 21.

Once we have observed the number of swimmers and that day's temperature over a period of several days, we can use those data to write a precise mathematical equation that describes the observed relationship. The equation is known as the prediction equation and is the basic component of regression analysis. Once you have written the prediction equation (using past observations), you can then insert any day's temperature into the equation. For purposes of our illustration, let's pick 105 degrees. The day's temperature is the independent variable. You could then solve the equation for 105 degrees and predict the value of the dependent variable. The dependent variable is the number of people seeking the cool water of the pool on a day having a temperature of 105 degrees. The techniques described also make it possible to determine the "goodness" of one's predictions. SPSS accomplishes this by applying the principles associated with inferential statistics, hypothesis testing, and significance levels. Therefore, we use the power of SPSS to calculate the prediction equation, use it to make predictions, and then determine whether the prediction is statistically significant.

OBJECTIVES

After completing this chapter, you will be able to

- Describe the purpose of single linear regression
- Input variable information and data for single linear regression
- Describe data assumptions required for single linear regression
- Interpret scatter plots concerning data assumptions for regression
- Interpret probability plots concerning data assumptions for regression
- Use SPSS to conduct single linear regression analysis
- Describe and interpret SPSS output for single linear regression
- Interpret the coefficients table for single linear regression
- Write the prediction equation and use SPSS to make predictions
- Write the prediction equation and use a calculator to make predictions

20.2 Research Scenario and Test Selection

The scenario used to explain the SPSS regression function centers on a continuation of the pool example presented in the introduction. You will enter hypothetical data for the number of patrons (dependent variable) at a public swimming pool and that day's temperature (independent variable). The research will investigate the relationship between a day's temperature and the number of patrons at the public swimming pool. The researcher also wishes to develop a way to estimate the number of patrons based on a day's temperature. It appears that single linear regression might be appropriate, but there are data requirements that must be met.

One data requirement (assumption) that must be met before using linear regression is that the distributions for the two variables must approximate the normal curve. There must also be a linear relationship between the variables. Also, the variances of the dependent variable must be equal for each level of the independent variable. The equality of variances is called homoscedasticity and is illustrated by a scatter plot that uses standardized residuals (error terms) and standardized prediction values. And yes, we must assume that the sample was random.

△ 20.3 RESEARCH QUESTION AND THE NULL HYPOTHESIS

The current research investigates the relationship between a day's temperature and the number of patrons at the swimming pool. We are interested in determining the strength and direction of any identified relationship between the two variables of temperature and number of patrons. If possible, we wish to develop a reliable prediction equation that would estimate the number of patrons on a day having a temperature that was not directly observed. We also wish to generalize to other days having the same temperature and to specify the number of expected patrons on those days.

The researcher wishes to better understand the influence that daily temperature may have on the number of public pool patrons. The alternative hypothesis is that the daily temperature directly influences the number of patrons at the public pool. The null hypothesis is the opposite; the temperature has no influence on the number of pool patrons.

△ 20.4 DATA INPUT

In this section, we enter the hypothetical data for 30 summer days, selected at random, from the records of a public swimming pool in a midwestern city. Patrons were counted via ticket sales, whereas temperature was recorded at noon on each day. These observations are then analyzed using single linear regression procedures provided in SPSS. As in the recent chapters, you are not given detailed instructions on entering the variable information and the data—most likely you are quite proficient at this by now. Begin by entering the variable information into the Variable View page as depicted in Figure 20.1.

Figure 20.1 Appearance of the Variable View Page After Entering Variable Information

swim.sav [DataSet1] - PASW Statistics Data Editor

File Edit View Data Transform Analyze Graphs Utilities Add-ons Window Help

	Name	Type	Width	Decimals	Label	Values	Missing	Columns	Align	Measure
1	temp	Numeric	8	0	Temperature at Noon (Fahrenheit)	None	None	8	Right	Scale
2	patrons	Numeric	8	0	Number of Patrons	None	None	8	Right	Scale

Next, enter the values into the Data View page for the two variables as presented in Figure 20.2. And remember, it is unnecessary to enter the Day# shown in Figure 20.2 as these are automatically entered as SPSS case numbers.

Figure 20.2 Daily Temperature and Patron Data for New SPSS Database Called Swim

Day#	Temp	Patrons	Day#	Temp	Patrons	Day#	Temp	Patrons
1	90	245	11	94	235	21	95	264
2	85	200	12	94	225	22	98	280
3	95	250	13	97	255	23	104	294
4	82	199	14	87	209	24	110	301
5	98	260	15	79	189	25	107	296
6	105	295	16	80	210	26	98	278
7	78	183	17	84	198	27	97	286
8	86	205	18	82	212	28	95	270
9	89	199	19	81	200	29	90	243
10	87	205	20	86	208	30	87	210

Let's begin with the bullet points that lead you through variable information input and data input (both procedures are greatly abbreviated) and then to the actual regression analysis. The interpretation of the output is presented at various stages of the analysis.

- Start SPSS.
- Click **Cancel** in the SPSS Statistics opening window.
- Click **File**, select **New**, and click **Data**.

- Click **Variable View** (enter all variable information as presented in Figure 20.1).
- Click **Data View** (carefully enter all data for the two variables as given in Figure 20.2) (you do not enter the Day# as it is done automatically by SPSS as case numbers).
- Click **File**, then click **Save As**, type **swim** in the File Name box, and then click **OK**.

You have now entered and saved 30 days of data for the number of swimmers and that day's temperature. Next, we check the distributions for normality.

△ 20.5 DATA ASSUMPTIONS (NORMALITY)

At this point, it is a good practice to test the distributions for normality, which is easily accomplished by following the bullet points given next. It will take a few clicks; it may seem like more than a few, but with very little practice it easily becomes routine.

- Click **Analyze**, select **Nonparametric Tests**, and then click **One-Sample** (the "One-Sample Nonparametric Tests" window opens—you have seen it before; it has three tabs at the top: Objective, Fields, and Settings).
- Click the **Objective** tab, and then click **Customize analysis**.
- Click the **Fields** tab (if your two variables are not in the Test Fields pane, then move them).
- Click the **Settings** tab, click **Customize tests**, and then click the **Kolmogorov-Smirnov test**.
- Click **Run** (the Output Viewer opens showing Figure 20.3).

Figure 20.3 Kolmogorov-Smirnov Test Results Showing Both Distributions Are Normal

Hypothesis Test Summary

	Null Hypothesis	Test	Sig.	Decision
1	The distribution of Temperature at Noon (Fahrenheit) is normal with mean 91.333 and standard deviation 8.608.	One-Sample Kolmogorov-Smirnov Test	.728	Retain the null hypothesis.
2	The distribution of Number of Patrons is normal with mean 236.8 and standard deviation 37.752.	One-Sample Kolmogorov-Smirnov Test	.138	Retain the null hypothesis.

Asymptotic significances are displayed. The significance level is .05.

The results of the Kolmogorov-Smirnov test (think of Vodka although, in my opinion, Scotch is better) are satisfactory as both distributions are determined to be normal. Now that SPSS has provided evidence that the variables approximate a normal distribution, we may move forward with our regression analysis.

20.6 REGRESSION AND PREDICTION △

When using SPSS to conduct regression analysis, it is easy to become overwhelmed with the number of options available for analysis and also with the resulting output. We introduce just those options that will enable you to do basic regression analysis. As you become more proficient, you will go far beyond what is offered here.

Let's request that SPSS check for linearity, equal variances, and residuals (error terms) for normality. Assuming that the swim database is open, follow the procedure presented next. On completion of the analytic requests in the following bullet points, the output is presented in the following sections.

As mentioned earlier, once the procedure is completed, a large volume of output is produced. We explain the output tables and graphs one at a time. Some will be explained out of order of appearance. So please pay careful attention, and you should have no difficulty interpreting the output you generate.

- Click **Analyze**, select **Regression**, and then click **Linear**.
- Click **Number of Patrons**, and then click the arrow next to the Dependent: box.
- Click **Temperature at Noon**, and then click the arrow next to the Independent(s): box (at this point your screen should look like Figure 20.4).

Figure 20.4 Linear Regression Window for Swim Data Analysis

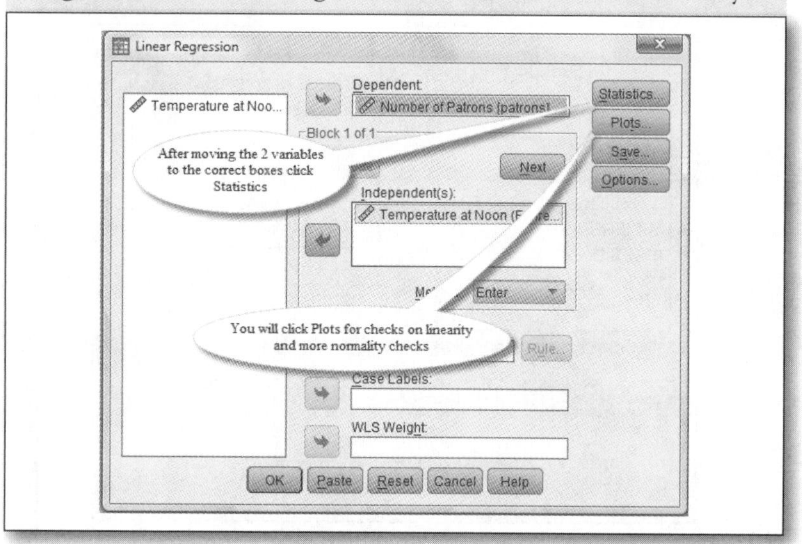

- Click **Statistics** (the "Linear Regression: Statistics" window opens; see Figure 20.5) (make sure that **Estimates** and **Model fit** are checked as shown).

Figure 20.5 "Linear Regression: Statistics" Window

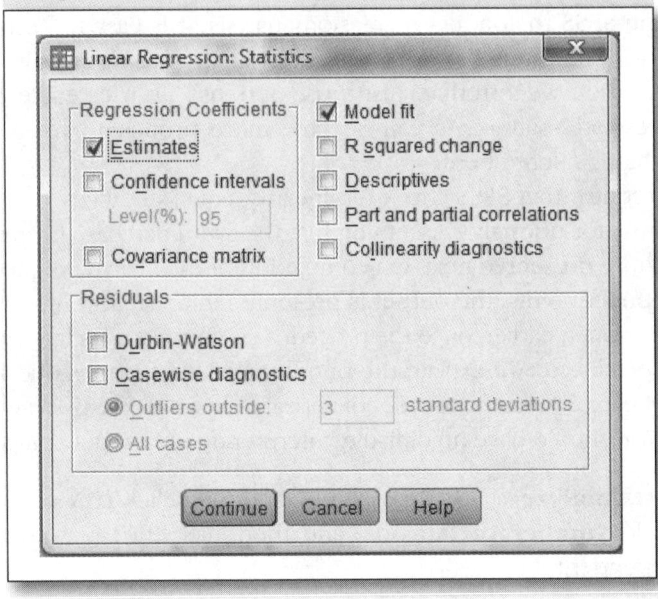

- Click **Continue**.

Figure 20.6 Plots Window: Further Data Assumption Checks

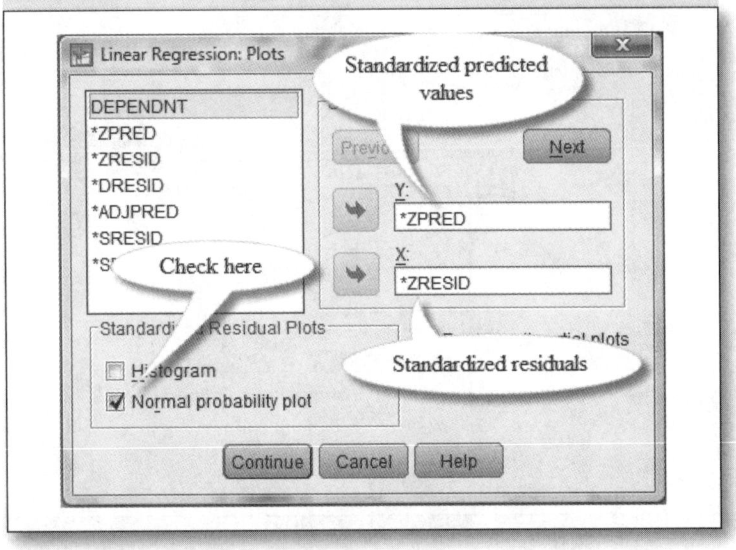

- Click **Plots** (the "Linear Regression: Plots" window opens; see Figure 20.6).
- Click ***ZPRED**, and then click the arrow next to the Y: box.
- Click ***ZRESID**, and then click the arrow next to the X: box.
- Click **Normal probability plot**.
- Click **Continue**, and then click **OK**.

Once you click **OK**, the Output Viewer opens with five tables and two graphs—please don't be dismayed. We will interpret the relevant aspects of the output in a careful and systematic manner.

20.7 INTERPRETATION OF OUTPUT (DATA ASSUMPTIONS) △

We continue our analysis and interpretation of the SPSS output by first looking at the tests intended to confirm that additional data requirements for regression are satisfied. Let's look at the three graphs that can be found in the Output Viewer. Hopefully, on examination of these graphs we will have more confidence that the data meet the assumptions required when choosing the linear regression procedure.

You may recall that you checked "Normal probability plot" in the "Linear Regression: Plots" window as shown in Figure 20.6. The result of this request

Figure 20.7 Normal P-P Plot: Standardized Residuals (Error Terms)

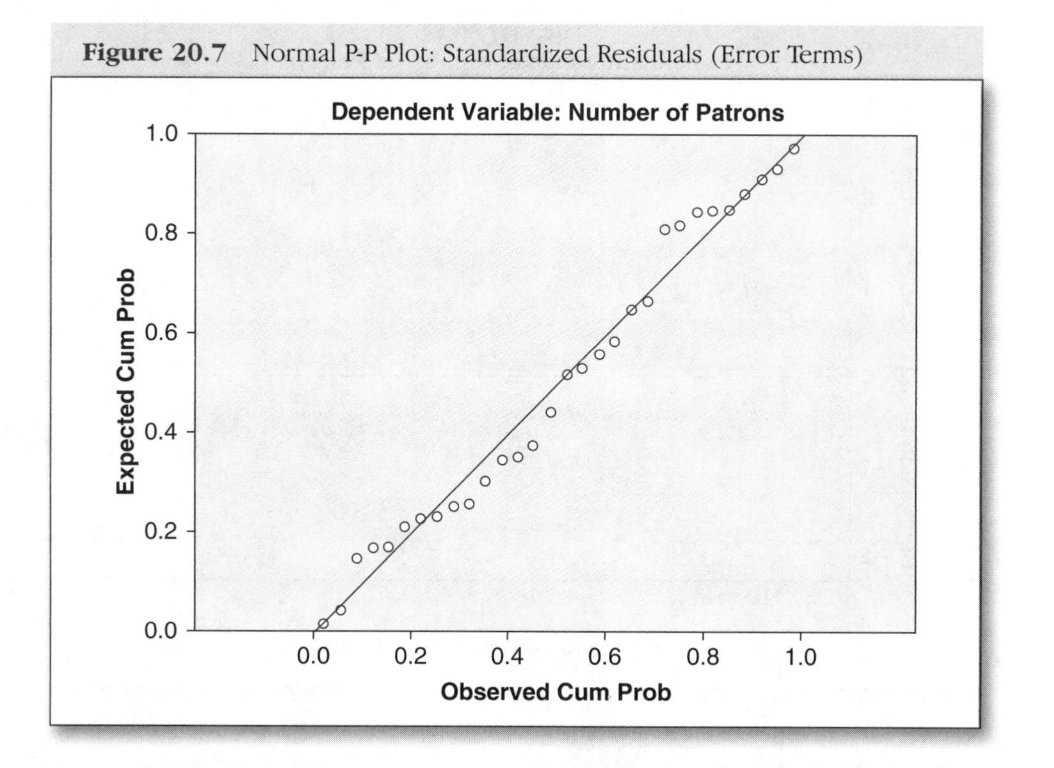

is shown in Figure 20.7. The observation that the small circles are close to the straight line provides evidence that the error terms (difference between predicted and observed values) are indeed normally distributed. If the small circles strayed far from the line, then we must assume non-normal data and seek statistical procedures other than linear regression.

The final plot, Figure 20.8, results from the requests we made as shown in Figure 20.6. The scatter plot combines the standardized predicted values (*ZPRED) with the values for the standardized residuals (*ZRESID). For our regression to be accurate, there must be equal variability in the dependent variable for each level of the independent variable. Note that the plot you produce will not have the line titled "best fit line" as shown in Figure 20.8. We have added the line to help you visualize how the bivariate variances are distributed along the best-fit line. These points represent the standardized residuals and the standardized predicted values. Looking at Figure 20.8, we see that the small circles are randomly dispersed with a corresponding lack of pattern. Such an appearance indicates equal variances in the dependent variable for each level of the independent variable. This data characteristic is referred to as homoscedasticity and is a requirement for using our linear regression. We can say that this requirement of linear regression is satisfied and continue with our analysis.

Figure 20.8 Standardized Residuals and Standardized Predicted Values

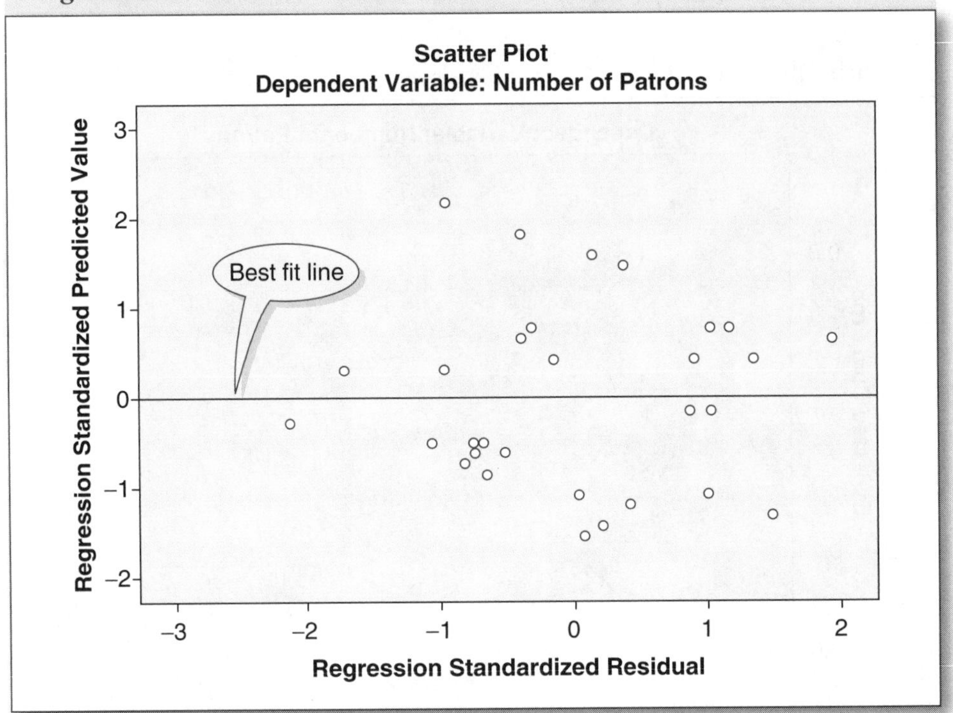

Now that all the data assumptions required for using the single linear regression approach have been met, we interpret the actual regression analysis.

20.8 INTERPRETATION OF OUTPUT (REGRESSION AND PREDICTION) △

Let's begin our interpretation of the regression output with the table titled Model Summary (shown in Figure 20.9) and see what can be learned about the relationship between our two variables. Most of the information in the model summary deals with the strength of the relationship between the dependent variable and the model.

Figure 20.9 Model Summary for the Number of Patrons (Dependent Variable)

Model Summary[b]

Model	R	R Square	Adjusted R Square	Std. Error of the Estimate	Change Statistics				
					R Square Change	F Change	df1	df2	Sig. F Change
1	.938[a]	.880	.875	13.334	.880	204.480	1	28	.000

a. Predictors: (Constant), Temperature at Noon (Fahrenheit)

b. Dependent Variable: Number of Patrons

The "R" is the correlation coefficient between the two variables; in this case, the correlation between temperature and number of patrons is high at .938. The next column, "R Square," indicates the amount of change in the dependent variable that can be attributed to our one independent variable. The R Square of .880 indicates that 88% (100 × .880) of the variance in the number of patrons (dependent variable) can be explained by the temperature of the day (independent variable). We now begin to conclude that we have a "good" predictor for number of expected patrons when consideration is given to the day's temperature. Next we examine the ANOVA table shown in Figure 20.10.

The ANOVA table, presented in Figure 20.10, indicates that the model can accurately explain variation in the dependent variable. We are able to say the model "accurately" explains variation, since the significance value of .000 informs us that the probability is very low that the variation explained by the model is due to chance. The conclusion is that changes in the dependent variable resulted from changes in the independent variable(s). In this example, changes in daily temperature resulted in significant changes in number of pool patrons.

Figure 20.10 ANOVA Table Indicating a Significant Relationship Between the Variables

ANOVA[b]

Model		Sum of Squares	df	Mean Square	F	Sig.
1	Regression	36352.890	1	36352.890	204.480	.000[a]
	Residual	4977.910	28	177.782		
	Total	41330.800	29			

a. Predictors: (Constant), Temperature at Noon (Fahrenheit)

b. Dependent Variable: Number of Patrons

Prediction

Now comes an interesting (and for some exciting) part of the analysis, where the Compute Variable function of SPSS and the regression output are used to define the prediction equation. We will predict unknown values based on past observations.

Figure 20.11 Coefficient Values for the Prediction Equation

Coefficients[a]

Model		Unstandardized Coefficients		Standardized Coefficients	t	Sig.
		B	Std. Error	Beta		
1	(Constant)	-138.877	26.384		-5.264	.000
	Temperature at Noon (Fahrenheit)	4.113	.288	.938	14.300	.000

a. Dependent Variable: Number of Patrons

The coefficients table presented in Figure 20.11 is most important when writing and using the prediction equation. Please don't glaze over, but we must present some basic statistics before you can use SPSS to do the hard work involved in prediction. The prediction equation takes the following form:

$$\hat{y} = a + bx,$$

where $\hat{y}$ = the predicted value, a = the intercept, b = the slope, and x = the independent variable.

Let's quickly define a couple of terms in the prediction equation that you may not be familiar with. The slope (b) records the amount of change in the dependent variable (number of patrons) when the independent variable (day's temperature) increases by one unit. The intercept (a) is the value of the dependent variable when $x = 0$.

In simple words, the prediction equation states that you multiply the slope (b) by the values of the independent variable of temperature (x) and then add the result of the multiplication (bx) to the intercept (a)—not too difficult. But where do you find the values for the intercept and the slope? Figure 20.11 provides the answers; The Constant is the intercept (a) and Temperature at Noon is the slope (b). The x values are already recorded in the database as "temp"—you now have everything required to solve the equation and make predictions. Substituting the regression coefficients, the slope and intercept, into the equation, we find the following:

$$\hat{y} = -138.877 + (4.113x)$$

The x represents any day's temperature that might be of interest.

Let's put SPSS to work and use our new prediction equation to make predictions for all the observed temperature values. By looking at the observed numbers of patrons and the predicted number of patrons, we can see how well the equation performs.

The following procedure assumes that SPSS is running and that the swim database is open.

- Click **Transform**, and then click **Compute Variable** (the "Compute Variable" window opens; see Figure 20.12, which shows the upper portion of the window with the completed operations as described in the following bullet points).
- Click the **Target Variable** box, and then type **pred_patrons**.
- Click the **Type and Label** box (the "Compute Variable: Type and Label" window opens), and then type **predicted number of patrons** in the box titled Label.
- Click **Continue**.
- Click the **Numeric Expression** box and then type **–138.877 + (4.113*temp)** (if you prefer, you can use the keypad in the "Compute Variable" window and the variable list to write this expression).
- Click **OK** at the bottom of the "Compute Variable" window (not shown in Figure 20.12).

Figure 20.12 Upper Portion of the Completed Compute Variable Window

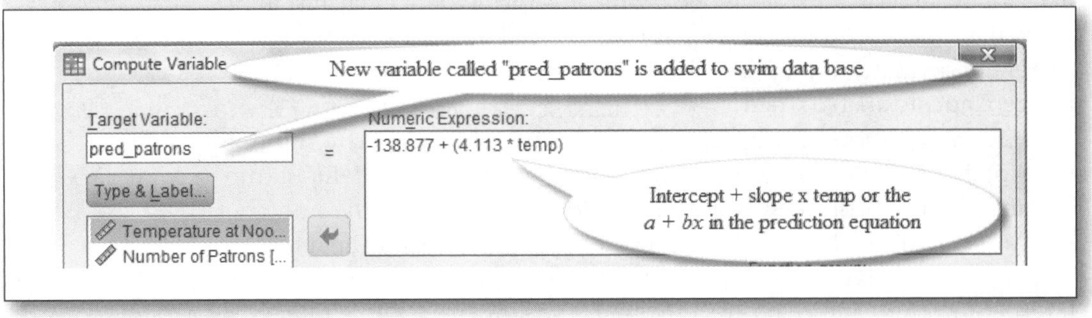

Once you click **OK**, SPSS creates a variable called **pred_patrons** and automatically inserts the new variable into your swim database. If the database does not appear on your screen, then do the following:

- Click **swim.sav** at the bottom of your screen, which will open that database (you will see the new variable added to your database, a portion of which is shown in Figure 20.13).

Reading these predicted values in conjunction with observed daily temperatures informs us as to what the equation would predict for each daily temperature. An example would be to look at Case 3, where the day's temperature was 95 degrees, 250 patrons were observed, and the equation predicted 251.86, or 252. You would be safe to use the 252 as an estimate of the number of patrons expected on days having a temperature of 95 degrees. When using the equation, you must, of course, consider any intervening circumstances such as thunderstorms and holidays.

There are also other uses for the prediction equation, such as inserting a value into the equation that was not directly observed and then using the

Figure 20.13 Swim Database With New Variable for Predicted Values

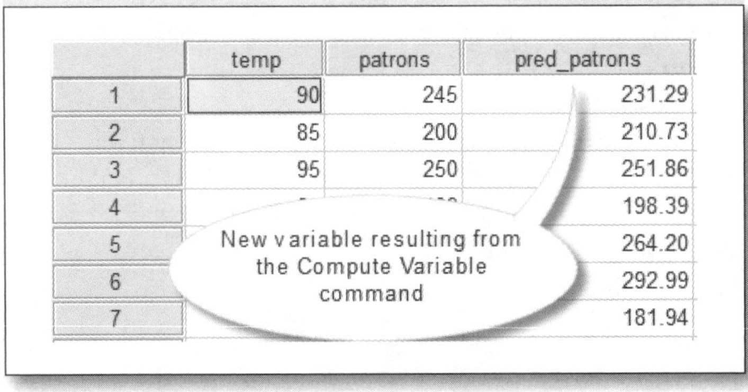

	temp	patrons	pred_patrons
1	90	245	231.29
2	85	200	210.73
3	95	250	251.86
4			198.39
5			264.20
6			292.99
7			181.94

equation to predict the number of patrons. Briefly, you insert the unobserved value (let's say 83 degrees) into the expression in the Numeric Expression box found in Figure 20.12 (replace "temp" with "83"), create a new variable, and then let SPSS solve the equation. You could repeat the process for any unknown x values that might be useful in your study. Or you could choose to do it by handheld calculator as follows. Using $\hat{y} = a + bx$ or $\hat{y} = -138.877 + 4.113(83)$, the answer is $\hat{y} = 202.5$.

20.9 RESEARCH QUESTION ANSWERED △

The single linear regression was chosen to investigate whether the number of swimming pool patrons was influenced by the day's temperature. The alternative hypothesis was that there was a relationship because the number of patrons was influenced by the daily temperature. The null hypothesis was the opposite, that daily temperature did not affect the number of swimming pool patrons. How well did our single regression answer these questions? Was the null hypothesis rejected? Was it possible to generate any statistical support for our alternative hypothesis?

We were able to support the proposition that single linear regression was applicable to our research question since data assumptions were met through statistical analysis (see Figures 20.3, 20.7, and 20.8). The developed prediction equation (the model) was discovered to reduce the error in predicting the number of pool patrons by 88% (see Figure 20.9). The significant F of .000 provided evidence that there was an extremely low probability (less than .0005) that the daily temperature explanation of the variation in the number of pool patrons was the result of chance (see Figure 20.10). Empirical evidence was also supportive when the prediction was used to calculate predicted values and then compared with the actual observations (see Figure 20.13).

20.10 SUMMARY △

This chapter presented single linear regression analyses. With single linear regression, you have a single independent variable and one dependent variable. The object of the analysis is to develop a prediction equation that permits the estimation of the dependent variable based on knowledge of the single independent variable. In the following chapter, we extend the concept of one independent variable to multiple independent variables and call it multiple. Multiple linear regression results when you have two or more independent variables when attempting to predict the value of the dependent variable.

Chapter 21

Multiple Linear Regression

△ 21.1 Introduction and Objectives

In the previous chapter, we covered single linear regression. The single regression approach is used when you have one independent variable and one dependent variable. This chapter presents multiple linear regression, which is used when you have two or more independent variables and one dependent variable. The research question for those using multiple regression concerns how the multiple independent variables, either by themselves or together, influence changes in the dependent variable. You use the same basic concepts as with single regression, except now you have multiple independent variables. The object of multiple linear regression is to develop a prediction equation that permits the estimation of the dependent variable based on knowledge of the independent variables.

The data requirements for multiple linear regression are the same as with single linear regression. These assumptions are discussed in Section 21.2. In keeping with the premise that this text is a primer, the multiple regression example used is as basic as possible. By basic we mean that the example presented has just two independent variables, the minimum number required when using multiple regression. We ask the reader to be cognizant of the fact that more independent variables could be analyzed using the same techniques described in this chapter.

OBJECTIVES

After completing this chapter, you will be able to

- Describe the purpose of multiple linear regression
- Input variable information and data for multiple linear regression
- Describe data assumptions required for multiple linear regression
- Use SPSS to conduct multiple linear regression analysis
- Interpret scatter plots concerning data assumptions for regression
- Interpret probability plots concerning data assumptions for regression
- Describe and interpret SPSS output from multiple linear regression
- Interpret ANOVA analysis as it relates to multiple linear regression
- Interpret the coefficients table for multiple linear regression
- Write the prediction equation and use it to make predictions
- Write the prediction equation and use a calculator to make predictions

21.2 RESEARCH SCENARIO AND TEST SELECTION △

The researcher wants to understand how certain physical factors may affect an individual's weight. The research scenario centers on the belief that height and age (independent variables) are related to an individual's weight (dependent variable). Another way of stating the scenario is that age and height influence the weight of an individual. When attempting to select the analytic approach, an important consideration is the level of measurement. As with single regression, the dependent variable must be measured at the scale level (interval or ratio). In this example, all data are measured at the scale level. What type of statistical analysis would you suggest to answer these questions?

Regression analysis comes to mind since we are attempting to estimate (predict) the value of one variable based on the knowledge of others, which can be done with a prediction equation. Single regression can be ruled out since we have two independent variables and one dependent variable. Let's consider multiple linear regression as a possible analytic approach.

We must check to see if our variables are approximately normally distributed. Furthermore, it is required that the relationship between the variables be approximately linear. And we will also check for homoscedasticity, which means that the variances in the dependent variable are the same for each level of the independent variables. Here's an example for homoscedasticity. A distribution of individuals 61 inches tall and aged 41 years would have the same variability in weight as those 72 inches tall and aged 31 years. In sections that follow, some of these required data characteristics will be examined immediately, others when we get deeper into the analysis.

△ 21.3 Research Question and the Null Hypothesis

The basic research question (alternative hypothesis) is whether an individual's weight is related to that person's age and height. The null hypothesis is the opposite of the alternative hypothesis: An individual's weight is not related to his or her age and height.

Therefore, this research question involves two independent variables, height and age, and one dependent variable, weight. The investigator wishes to determine how height and age, taken together or individually, might explain the variation in weight. Such information could assist someone attempting to estimate an individual's weight based on the knowledge of his or her height and age. Another way of stating the question uses the concept of prediction and error reduction. How successfully could we predict someone's weight given that we know his or her age and height? How much error could be reduced in making the prediction when age and height are known? One final question: Are the relationships between weight and each of the two independent variables statistically significant?

△ 21.4 Data Input

In this section, you enter hypothetical data for 12 randomly selected individuals measured on weight, height, and age. You then use SPSS to analyze these data using multiple linear regression. As in recent chapters, detailed instructions on entering the variable information and the data are not given. The Variable View page, shown in Figure 21.1, serves as a guide for the entry of variable information.

Figure 21.1 Variable View for Multiple Regression Data: Three Variables

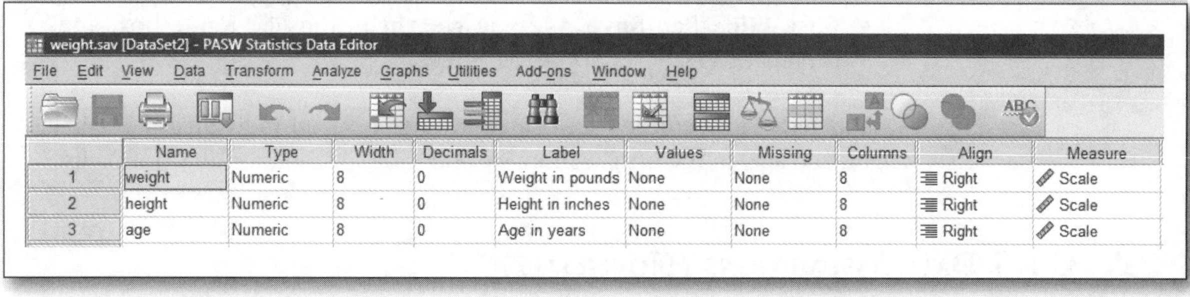

weight.sav [DataSet2] - PASW Statistics Data Editor

File Edit View Data Transform Analyze Graphs Utilities Add-ons Window Help

	Name	Type	Width	Decimals	Label	Values	Missing	Columns	Align	Measure
1	weight	Numeric	8	0	Weight in pounds	None	None	8	Right	Scale
2	height	Numeric	8	0	Height in inches	None	None	8	Right	Scale
3	age	Numeric	8	0	Age in years	None	None	8	Right	Scale

Figure 21.2 contains the data for the 12 individuals. It is not necessary to enter the Case# into the SPSS Data View page.

Figure 21.2 Data for Multiple Linear Regression

Case#	Weight	Height	Age
1	115	62	41
2	140	62	21
3	125	62	31
4	125	64	21
5	145	64	31
6	135	64	41
7	165	72	41
8	190	72	31
9	175	72	21
10	150	66	31
11	155	66	31
12	140	64	21

Follow the procedure below and enter both the variable information and the data for the three variables, and save the file as instructed.

- Start SPSS.
- Click **Cancel** in the SPSS Statistics opening window.
- Click **File**, select **New**, and click **Data**.
- Click **Variable View** (enter all variable information as presented in Figure 21.1).

- Click **Data View** (carefully enter all data for Weight, Height, and Age given in Figure 21.2) (do not enter the Case#).
- Click **File**, then **Save As**, type **weight** in the File Name box, and then click **OK**.

You have now entered and saved the data for an individual's weight, height, and age. In the next section, we check the distributions for normality.

Δ 21.5 DATA ASSUMPTIONS (NORMALITY)

As was done in the previous chapter, we first check the data distributions for normality.

- Click **Analyze**, select **Nonparametric Tests**, and then click **One-Sample** (the "One-Sample Nonparametric Tests" window opens).
- Click the **Objective** tab, and then click **Customize analysis**.
- Click the **Fields** tab (if your three variables are not in the Test Fields pane, then move them).
- Click the **Settings** tab, click **Customize tests**, and then click the **Kolmogorov-Smirnov test**.
- Click **Run** (the Output Viewer opens showing Figure 21.3).

Figure 21.3 Hypothesis Tests for Normality

Hypothesis Test Summary

	Null Hypothesis	Test	Sig.	Decision
1	The distribution of Weight in pounds is normal with mean 146.667 and standard deviation 21.881.	One-Sample Kolmogorov-Smirnov Test	.995	Retain the null hypothesis.
2	The distribution of Height in inches is normal with mean 65.833 and standard deviation 3.95.	One-Sample Kolmogorov-Smirnov Test	.382	Retain the null hypothesis.
3	The distribution of Age in years is normal with mean 30.167 and standard deviation 7.93.	One-Sample Kolmogorov-Smirnov Test	.668	Retain the null hypothesis.

Asymptotic significances are displayed. The significance level is .05.

The results of the Kolmogorov-Smirnov tests indicate that the three variables are indeed normally distributed—we may proceed with regression analysis.

21.6 REGRESSION AND PREDICTION Δ

As we did with the single linear regression in Chapter 20, we now check the data for linearity, equal variances, and normality of the error terms (residuals). If it's not already running, open **weight.sav** and follow the procedure presented next. Once you have completed all these analytic requests, you will see the output as presented in the following sections.

- Click **Analyze**, select **Regression**, and then click **Linear** (the "Linear Regression" window opens; see Figure 21.4 for its appearance after moving the variables).
- Click **Weight**, and then click the arrow next to the Dependent: box.
- Click **Height**, and then click the arrow next to the Independent(s): box.
- Click **Age**, and then click the arrow next to the Independent(s): box (at this point, your screen should look like Figure 21.4).

Figure 21.4 "Linear Regression" Window After Moving Variables

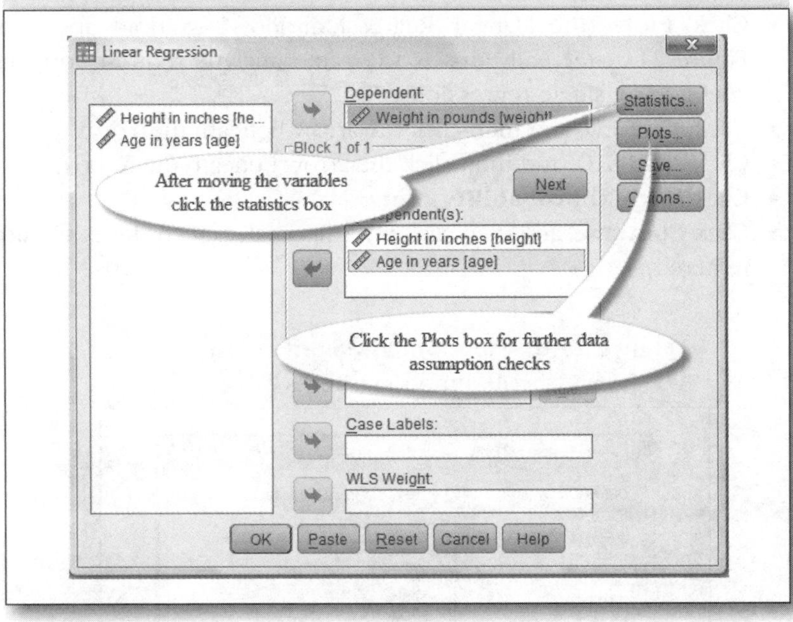

- Click **Statistics** (the "Linear Regression: Statistics" window opens; see Figure 21.5).
- Click **Estimates**, and then click **Model fit** (see Figure 21.5).
- Click **Continue** (returns to the "Linear Regression" window depicted in Figure 21.4).

Figure 21.5 "Linear Regression: Statistics" Window

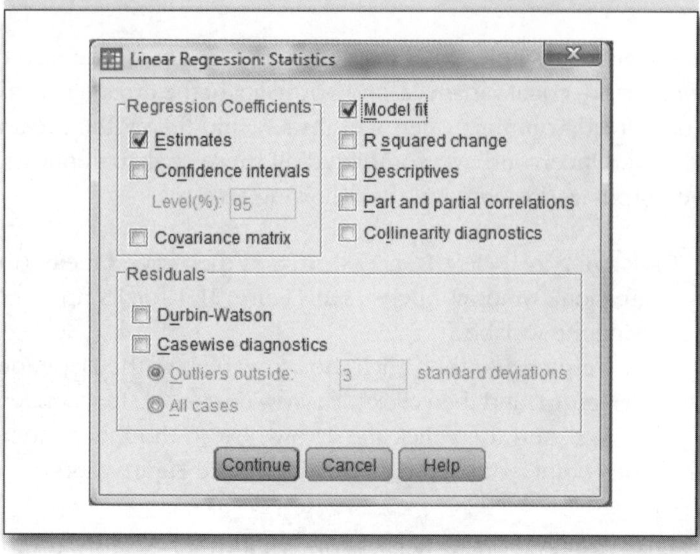

- Click **Plots** (the "Linear Regression: Plots" window opens; see Figure 21.6) (actually this is the same analytic request you made when doing single regression).
- Click ***ZPRED**, and then click the arrow beneath the Y: box.
- Click ***ZRESID**, and then click the arrow beneath the X: box.
- Click **Normal probability plot**.
- Click **Continue**, and then click **OK** (this final click produces all output required).

Figure 21.6 Plots Window: Further Data Assumption Checks

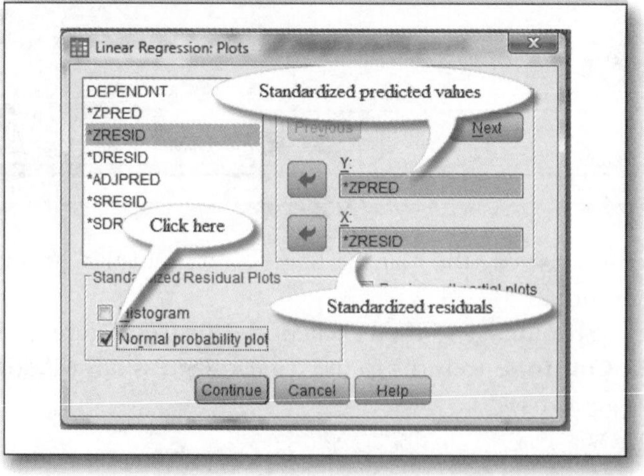

You now have all the output required to finalize and interpret additional data assumptions and your multiple linear regression analysis.

21.7 INTERPRETATION OF OUTPUT (DATA ASSUMPTIONS) △

Figure 21.6 shows that the "Normal probability plot" box was checked. The result is shown in Figure 21.7. The small circles are close to the straight line, which provides evidence that the residuals (error terms) are indeed normally distributed.

The final plot, Figure 21.8, results from the requests we made as shown in Figure 21.6. As with single variable regression, the scatter plot combines the standardized predicted values (*ZPRED) with the values for the standardized residuals (*ZRESID). Since the small circles follow no pattern—they are randomly dispersed in the scatter plot—we assume equality of variances. There are numerous appearances that Figure 21.8 may take on that would indicate unequal variances. One such appearance is referred to as the "bow tie" scatter plot. The "bow tie" scatter plot has the error terms

Figure 21.7 Normal P-P Plot Regression Standardized Residuals (Error Terms)

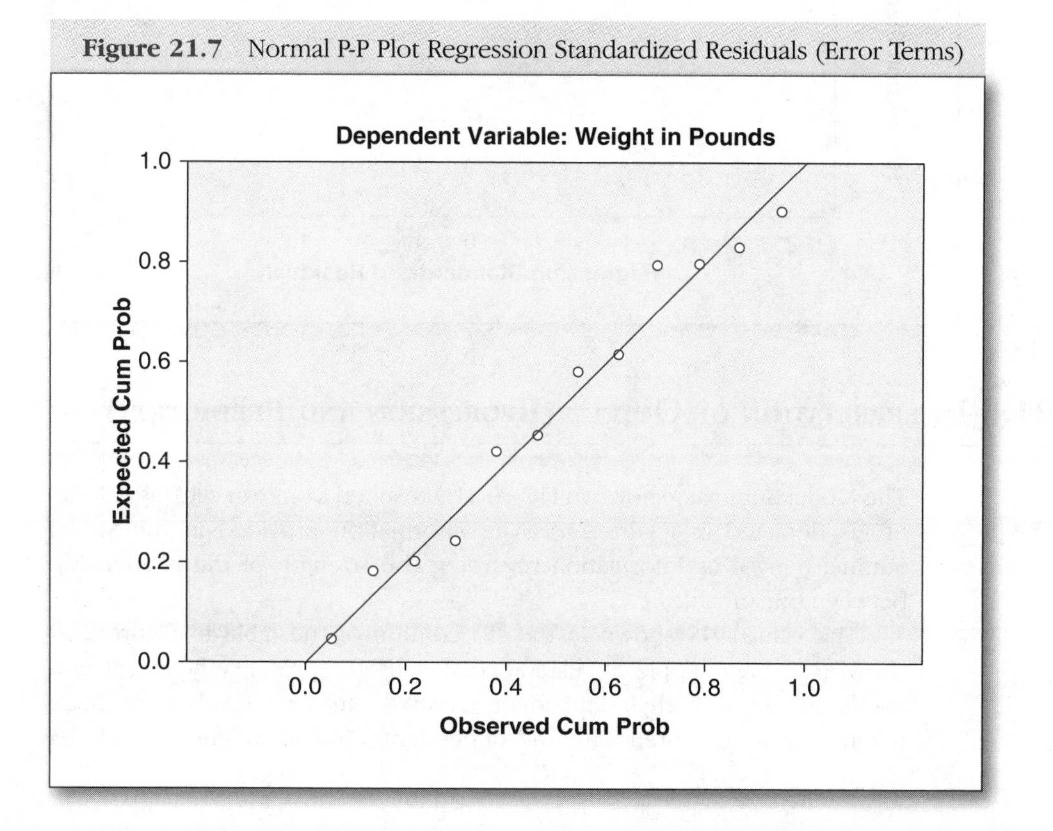

bunched up along both verticals and then tapering toward the middle, which is the zero point in Figure 21.8. There are many other shapes that indicate unequal variances.

Figure 21.8 Scatter Plot of Residuals: Lack of Pattern Indicates Equal Variances

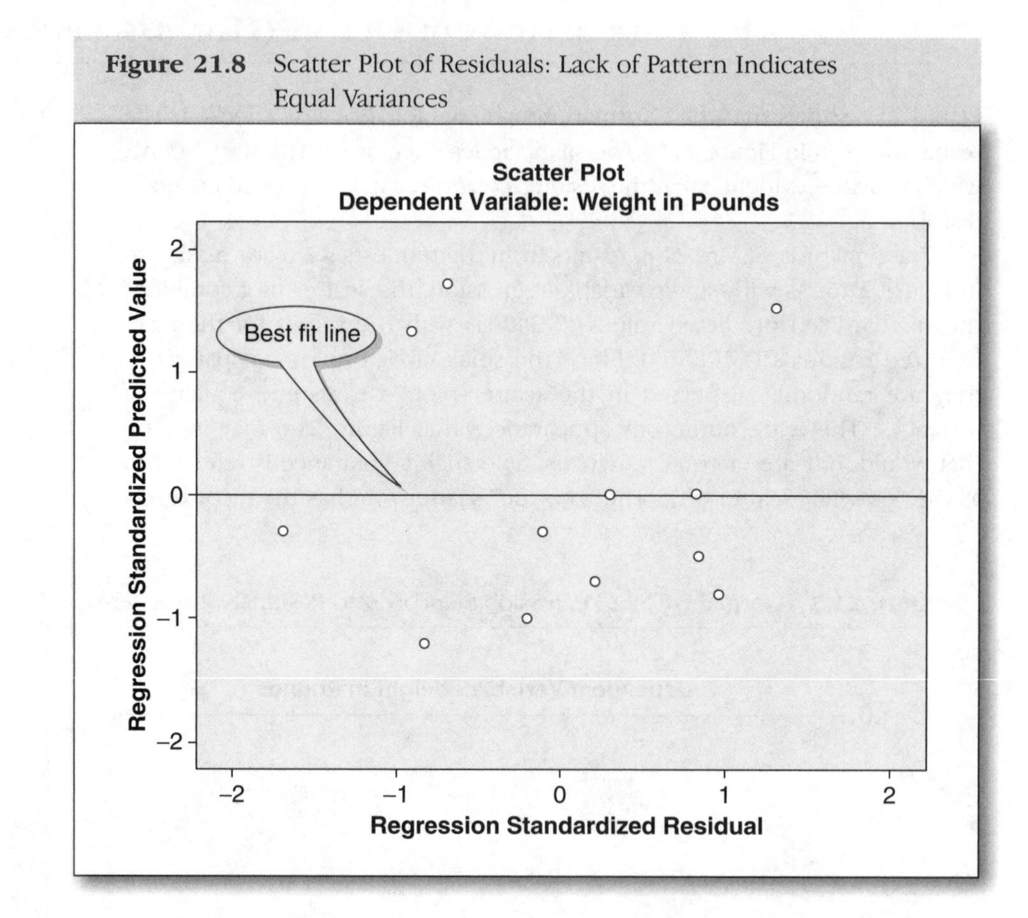

Δ 21.8 INTERPRETATION OF OUTPUT (REGRESSION AND PREDICTION)

The Model Summary shown in Figure 21.9 resulted from you clicking "Model fit" as depicted in Figure 21.5. The information provided in the Model Summary gives us information regarding the strength of the relationship between our variables.

The value .919 shown in the "R" column of the table in Figure 21.9 shows a strong multiple correlation coefficient. It represents the correlation coefficient when both independent variables (age and height) are taken together and compared with the dependent variable weight. The Model

Figure 21.9 Model Summary for Multiple Linear Regression

Model Summary[b]

Model	R	R Square	Adjusted R Square	Std. Error of the Estimate
1	.919[a]	.845	.811	9.515

a. Predictors: (Constant), Age in years, Height in inches

b. Dependent Variable: Weight in pounds

Summary indicates that the amount of change in the dependent variable is determined by the two independent variables—not by one as in single regression. From an "interpretation" standpoint, the value in the next column, "R Square," is extremely important. The R Square of .845 indicates that 84.5% (.845 × 100) of the variance in an individual's weight (dependent variable) can be explained by both the independent variables height and age. It is safe to say that we have a "good" predictor of weight if an individual's height and age are known. We next examine the ANOVA table shown in Figure 21.10.

The ANOVA table indicates that the mathematical model (regression equation) can accurately explain variation in the dependent variable. The value of .000 (less than .05) provides evidence that there is a low probability that the variation explained by the model is due to chance. We conclude that changes in the dependent variable result from changes in the independent variables. In this example, changes in height and age resulted in significant changes in weight.

Figure 21.10 ANOVA Table Indicating a Significant Relationship Between the Variables

ANOVA[b]

Model		Sum of Squares	df	Mean Square	F	Sig.
1	Regression	4451.886	2	2225.943	24.588	.000[a]
	Residual	814.780	9	90.531		
	Total	5266.667	11			

a. Predictors: (Constant), Age in years, Height in inches

b. Dependent Variable: Weight in pounds

Prediction

As in the prior single regression chapter, we come to that interesting point in regression analysis where we discover the unknown. We accomplish such "discovery" by using a prediction equation to make estimates based on our 12 original observations. Let's see how the Coefficients table in Figure 21.11 can make such discovery possible.

Figure 21.11 Coefficient Values for Use in the Prediction Equation

Coefficients[a]

Model		Unstandardized Coefficients		Standardized Coefficients	t	Sig.
		B	Std. Error	Beta		
1	(Constant)	-175.175	48.615		-3.603	.006
	Height in inches	5.072	.727	.916	6.974	.000
	Age in years	-.399	.362	-.145	-1.103	.299

a. Dependent Variable: Weight in pounds

As with single linear regression, the Coefficients table provides the essential values for the prediction equation. The prediction equation takes the following form:

$$\hat{y} = a + b_1x_1 + b_2x_2,$$

where $\hat{y}$ = the predicted value, a = the intercept, b_1 = the slope for height, x_1 = the independent variable of height, b_2 = the slope for age, and x_2 = the independent variable of age.

You may recall that in the previous chapter on single regression, we defined the slope (b) and the intercept (a). Those definitions are the same for the multiple regression equation except that we now have a slope for each of the independent variables.

The equation simply states that you multiply the individual slopes by the values of the independent variables and then add the products to the intercept—not too difficult. The slopes and intercepts can be found in the table shown in Figure 21.11. Look in the column labeled "B." The intercept (the value for a in the above equation) is located in the "(Constant)" row and is −175.175.

The value below this of 5.072 is the slope for height and below that is the value of −.399, the slope for age. The values for x are found in the **weight.sav** database. Substituting the regression coefficients, the slope and intercept, into the equation, we find the following:

$$\hat{y} = 172.175 + (5.072 * height) + (-.399 * age)$$

We are now ready to use the Compute Variable function of SPSS and make some predictions. The following bullet points assume that SPSS is running and that the weight database is open.

- Click **Transform**, and then click **Compute Variable** (the "Compute Variable" window opens; see Figure 21.12, which shows the upper portion of the window with the completed operations as described in the following bullet points).
- Click the **Target Variable** box, and then type **pred_weight**.
- Click the **Type and Label** box, and a window opens; then type **predicted weight** in the Label box.
- Click **Continue** (the "Compute Variable" window remains as shown in Figure 21.12).
- Click the **Numeric Expression** box and type **−175.175** + **(5.072 * height)** + **(−.399 * age)** (if preferred, you could use the keypad in the "Compute Variable" window and the variable list to write this expression—try it, you may find it easier).
- Click **OK** at the bottom of this window (not shown in Figure 21.12).

Figure 21.12 Completed "Compute Variable" Window

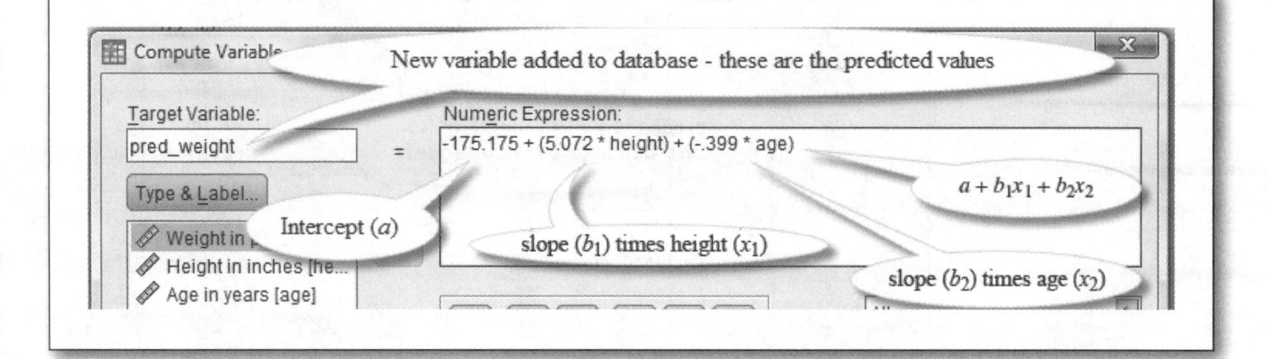

The new variable **pred_weight** is automatically inserted into the weight database once you click **OK** in the above series of bullet points. If the database does not appear on your computer screen, then do the following:

- Click **weight.sav** at the bottom of your screen, which will open that database (you will see the new variable added to your database, a portion of which is shown in Figure 21.13).

Figure 21.13 shows the new Data View screen with the just created variable. Let's look at Case 10 and interpret the newly created variable. Case 10 shows that someone measured at 66 inches in height and aged 31 weighed 150 pounds. The prediction equation estimated that an individual possessing those two values (66 and 31) would weigh 147.21 pounds. If you read each of the cases for height and age, you can read the actual observed weight as well as the prediction.

As we did for single regression, any values (for x_1 and x_2) that might be of special interest could be plugged into the equation and then solved for $\hat{y}$ (the predicted y value). One note of caution is that the values chosen should be within the range of the original observations to maintain accuracy. You can accomplish such predictions by substituting values into the Numeric expression panel in the "Compute Variable" window and creating a new variable as described in the previous chapter. The other method is to use a handheld calculator as was also described in the previous chapter.

Figure 21.13 Data View Showing New Variable From Use of the Prediction Equation

	weight	height	age	pred_weight
1	115	62	41	122.93
2	140	62	21	130.91
3	125	62	[illegible]	126.92
4	[illegible]	[illegible]	[illegible]	141.05
5	[illegible]	[illegible]	[illegible]	137.06
6	[illegible]	[illegible]	[illegible]	133.07
7	165	[illegible]	41	173.65
8	190	72	31	177.64
9	175	72	21	181.63
10	150	66	31	147.21
11	155	66	31	147.21
12	140	64	21	141.05

New variable - these are the predicted values resulting from the Compute Variable request

21.9 Research Question Answered

At the beginning of this chapter, we stated that the purpose of the research was to investigate whether a person's weight is influenced by his or her age and height. You might also recall that the null hypothesis was that age and height had no influence on a person's weight. How well did our multiple regression analysis answer these questions? And could we reject the null hypothesis and thereby provide evidence in support of our alternative hypothesis?

First, our questions concerning the required data assumptions for using multiple regression were answered in the affirmative. It was determined that multiple linear regression could be used (see Figures 21.3, 21.7, and 21.8). Next, the prediction equation, which was developed from previous observations, was found to reduce the error in predicting weight by 84.5% (see Figure 21.9). Additional statistical evidence supporting the value of our prediction equation was provided with the finding of a significant *F* test. The significance was less than .05 indicating a low probability that the explanation of the variation in weight by using age and height was the result of chance (see Figure 21.10). Empirical evidence in support of the prediction equation was also observed. The Compute Variable function of SPSS was used to calculate predicted values that could directly be compared with our observations (see Figure 21.13).

21.10 Summary

In this chapter, we presented multiple linear regression. With multiple linear regression, you have two or more independent variables and one dependent variable. The object was to write a prediction equation that would permit the estimation of the value of the dependent variable based on knowledge of two or more independent variables. In the following two chapters, we use the chi-square distribution in the analysis of categorical data.

CHAPTER 22

CHI-SQUARE GOODNESS OF FIT

△ 22.1 INTRODUCTION AND OBJECTIVES

Chi-square, designated by the symbol χ^2, is a very popular nonparametric statistical test appropriate when data are in the form of frequency counts or percentages, or proportions that can be converted to frequencies. Chi-square is appropriate for nominal data and can be used to compare frequencies occurring in categories. "Goodness of fit" refers to the question of whether the observed frequencies match the theoretical values. Chi-square deals with discrete variables. For example, if you roll a fair die 900 times, you would expect, in terms of probability, 150 "ones," 150 "twos," 150 "threes," 150 "fours," 150 "fives," and 150 "sixes." The question is, how much divergence from these expected frequencies are you willing to accept before you decide the die is somehow rigged and not fair?

The chi-square goodness of fit test allows you to determine whether the observed frequencies for a single variable differ from what is expected by chance. There are occasions when we would not expect equal frequencies in an analysis. For example, if we randomly select samples of cats from the feline population in a certain city, we would probably not expect the same frequencies for each breed of cat. The chi-square goodness of fit test allows you to determine whether the category frequencies obtained in your samples differ from expected samples that are not equal.

When using chi-square, you choose a level of significance such as .05, apply the test, and interpret the outcome to determine if the results are such that they would be expected by chance. The formula for χ^2 is

$$\chi^2 = \sum_{i=1}^{n} \frac{(Oi - Ei)^2}{Ei},$$

where O_i is the observed frequency, E_i is the expected frequency, and n is the number of observations. The mathematics involved in computing χ^2 involves subtracting the expected value from the observed value, squaring the result, dividing by the expected value, and then adding all the cases. Clearly, the closer the expected values are to the observed values, the smaller the value of χ^2. The chi-square goodness of fit test allows you to determine whether the observed group frequencies for a single variable differ from what is expected by chance.

OBJECTIVES

After completing this chapter, you will be able to

- Describe the purpose of the chi-square goodness of fit statistical test
- Explain the computation of χ^2
- Use χ^2 to test hypotheses regarding frequencies
- Interpret the results of an SPSS chi-square goodness of fit statistical test
- Interpret the research question and null hypothesis

22.2 RESEARCH SCENARIO AND TEST SELECTION: FIRST EXAMPLE △

We are particularly fond of DOTS® candy made by the Tootsie Roll® company. We each enjoy different flavors, so we were interested in determining if the five flavors (strawberry, lemon, cherry, orange, and lime) are equally distributed. In short, do the boxes contain the same number of each flavor? The chi-square goodness of fit test is appropriate to investigate this question because we are dealing with nominal data and frequencies of occurrence of each flavor of candy. Assumptions related to the chi-square test are as follows: (1) observations must be independent, (2) sample size is relatively large, such that the expected frequencies for each category should be at least 1, and the expected frequencies should be at least 5 for 80% or more of the categories.

22.3 RESEARCH QUESTION AND NULL HYPOTHESIS: FIRST EXAMPLE

A quick inspection of several boxes of the candy led us (the researchers) to question the belief that different flavors of candies would usually be distributed equally in the boxes if there were not a compelling reason to do otherwise. Perhaps the company did a marketing study and found that most people prefer a particular flavor.

We state the null and alternate hypotheses:

H_0: There is no difference in the frequency of each flavor of candy (the number of candies of each flavor are equal).

H_A: There is a difference in the frequency of each flavor of candy (the number of candies of each flavor is not equal).

22.4 DATA INPUT, ANALYSIS, AND INTERPRETATION OF OUTPUT: FIRST EXAMPLE

We purchased 13 boxes of DOTS from various stores in the area over a 4-month period of time, and we counted the total number of pieces in each box and tallied the number of flavors. A summary of our results is shown in Figure 22.1.

Figure 22.1 Frequency of the Five Flavors

Orange	Strawberry	Cherry	Lime	Lemon
131	117	263	147	123

What is the expected value for each flavor of candy? Since the total number of pieces of candy is 781, and we assume an equal number for the five flavors, the expected value for each flavor is 781/5 = 156.2.

We ask you, the reader, to perform a chi-square test, using the .05 level of significance, to determine if the flavors are equally distributed. Is the number of each the same? We will take you through the process step-by-step.

First, categorize the flavors as follows: 1 = Orange, 2 = Strawberry, 3 = Cherry, 4 = Lime, and 5 = Lemon. Now follow these steps:

- Start SPSS.
- Enter the following two variables in the Variable View screen: **Flavor** and **Frequencies**. For both variables, set type to numeric, decimals to 0, and measure to nominal.
- Switch to the Data View screen.
- Enter **1** in the cell below Flavor, and enter **131** in the cell below Frequencies.
- Enter **2** in the second cell below Flavor, and enter **117** in the second cell below Frequencies.
- Enter **3** in the third cell below Flavor, and enter **263** in the third cell below Frequencies.
- Enter **4** in the fourth cell below Flavor, and enter **147** in the fourth cell below Frequencies.
- Enter **5** in the fifth cell below Flavor, and enter **123** in the fifth cell below Frequencies.

Be certain to save the file as Flavor.

Now that you have entered the data, you may proceed with the analysis.

- Click **Data** on the Main menu, and then click **Weight Cases**.

A window titled Weight Cases will open.

- Click **Weight Cases by**, and then click **Frequencies** in the left panel and click the right arrow to move it to the Frequency Variable box.
- Click **OK**.
- Click **Analyze**, select **Nonparametric Tests,** select **Legacy Dialogs**, and then click **Chi-Square**. A window titled Chi-Square Test will open.
- Click **Flavo**r in the left panel, and click the right arrow to place it in the Test Variable List box.
- Click **All categories equal**.
- Click **OK**.

Figure 22.2 shows the observed and expected frequencies of each flavor and the differences between the observed and expected frequencies.

Figure 22.2 Observed and Expected Frequencies for Each Flavor

Flavor

	Observed N	Expected N	Residual
1.00	131	156.2	-25.2
2.00	117	156.2	-39.2
3.00	263	156.2	106.8
4.00	147	156.2	-9.2
5.00	123	156.2	-33.2
Total	781		

Figure 22.3 shows the results of the chi-square test.

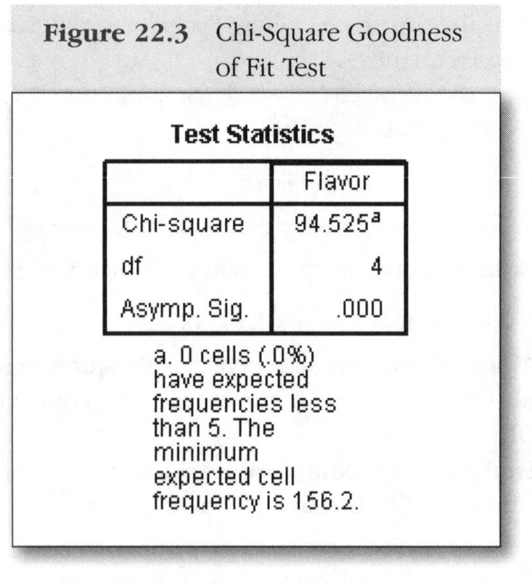

Figure 22.3 Chi-Square Goodness of Fit Test

Test Statistics

	Flavor
Chi-square	94.525[a]
df	4
Asymp. Sig.	.000

a. 0 cells (.0%) have expected frequencies less than 5. The minimum expected cell frequency is 156.2.

Although the table lists the significance level as .000, it does not mean there is no level of significance. SPSS rounds values, so the actual value is less than .0005, which SPSS rounds to .000.

The computed value of χ^2 is 94.53, and the level of significance is .000. Since .000 is less than .05, we reject the null hypothesis, which states there is no difference in the frequencies of flavors.

22.5 RESEARCH SCENARIO AND TEST SELECTION: SECOND EXAMPLE △

You have a die that you feel may be loaded. So you decide to test the die by throwing it 120 times and keeping track of the results (frequency of each face). The chi-square goodness of fit test is appropriate to investigate this question because we are dealing with nominal data and frequencies of occurrence of each face of the die.

22.6 RESEARCH QUESTION AND NULL HYPOTHESIS: SECOND EXAMPLE △

The researcher questions the honesty of this die because he has tossed it often and has noted that one face seems to turn up more frequently than might be expected. He states his null hypothesis and the alternative hypothesis.

H_0: There is no difference in the frequency of faces that show on the die (the frequencies of the six faces of the die are equal).

H_A: There is a difference in the frequency of faces that show on the die (the frequencies of the six faces of the die are not equal).

22.7 DATA INPUT, ANALYSIS, AND INTERPRETATION OF OUTPUT: SECOND EXAMPLE △

Let's set α at .05. Your results are shown in Figure 22.4.

Figure 22.4 Frequency of Each Face of a Die Thrown 120 Times

One	Two	Three	Four	Five	Six
13	28	16	10	32	21

- Start SPSS.
- Enter the following two variables in the Variable View screen: **Die** and **Frequencies**.
- Switch to the Data View screen.
- Enter **1** in the cell below Die, and enter **13** in the cell below Frequencies.

- Enter **2** in the second cell below Die, and enter **28** in the second cell below Frequencies.
- Continue entering data as shown in Figure 22.4.

Be certain to save the file as Die.

Now that you have entered the data, you may proceed with the analysis.

- Click **Data** on the Main menu, and then click **Weight Cases**.

A window titled Weight Cases will open.

- Click **Weight Cases by**, and then click **Frequencies** in the left panel, and click the right arrow to move it to the Frequency Variable box.
- Click **OK**.
- Click **Analyze**, select **Nonparametric Tests**, select **Legacy Dialogs**, and then click **Chi-Square**, and a window titled "Chi-Square Test" will open.
- Click **Die** in the left panel, and click the right arrow to place it in the Test Variable List box.
- Click **All categories equal**.
- Click **OK**.

Figure 22.5 shows the observed and expected frequencies for 120 tosses of the die and the differences between the observed and expected frequencies.

Figure 22.5 Observed and Expected Frequencies for 120 Tosses of the Die

Die

	Observed N	Expected N	Residual
1.00	13	20.0	-7.0
2.00	28	20.0	8.0
3.00	16	20.0	-4.0
4.00	10	20.0	-10.0
5.00	32	20.0	12.0
6.00	21	20.0	1.0
Total	120		

Figure 22.6 shows the results of the chi-square test.

Figure 22.6 Chi-Square Goodness of Fit Test

Test Statistics

	Die
Chi-square	18.700[a]
df	5
Asymp. Sig.	.002

a. 0 cells (.0%) have expected frequencies less than 5. The minimum expected cell frequency is 20.0.

The computed value of χ^2 is 18.70 and the level of significance is .002. Since .002 is less than .05, we reject the null hypothesis, which states there is no difference in the frequencies of the faces 1, 2, 3, 4, 5, and 6. We now have statistical evidence that the die is loaded.

22.8 SUMMARY

In this chapter, you learned how to apply the chi-square goodness of fit test to determine whether the observed occurrences of a single variable are consistent with expected frequencies. In the following chapter, you will apply chi-square to a situation involving two variables. The chi-square test of independence is a nonparametric test designed to determine whether two variables are independent or related.

CHAPTER 23

CHI-SQUARE TEST OF INDEPENDENCE

△ 23.1 INTRODUCTION AND OBJECTIVES

The chi-square test of independence is a nonparametric test designed to determine whether two variables are independent or related. This test is designed to be used with data that are expressed as frequencies. The discrete data to be analyzed are placed in a contingency table.

A contingency table is an arrangement in which a set of objects is classified according to two criteria of classification, one criterion being entered in rows and the other in columns. A contingency table may have more than two columns and more than two rows. To investigate the agreement between the observed and the expected frequencies, we compute the statistic

$$\chi^2 = \sum_{i=1}^{n} \frac{(\mathrm{O}i - \mathrm{E}i)^2}{\mathrm{E}i},$$

where the sum is taken over all cells in the contingency table (see Chapter 22).

OBJECTIVES

After completing this chapter, you will be able to

- Describe the purpose of the chi-square test of independence
- Explain the computation of χ^2
- Determine the degrees of freedom and expected values for a contingency table
- Use SPSS to conduct chi-square tests of independence
- Interpret the results of an SPSS test for chi-square

23.2 RESEARCH SCENARIO AND TEST SELECTION: FIRST EXAMPLE

A researcher had heard that color blindness is related to gender in certain populations. He wished to determine if this were true for a group of individuals for which he had collected data. Because the sample involves data measured at the nominal level and is concerned only with frequencies, the researcher decided that a chi-square test of independence would be appropriate. Assumptions related to the chi-square test are as follows: (1) observations must be independent, (2) sample size is relatively large, such that the expected frequencies for each category should be at least 1, and the expected frequencies should be at least 5 for 80% or more of the categories.

23.3 RESEARCH QUESTION AND NULL HYPOTHESIS: FIRST EXAMPLE

The researcher is aware that color blindness is a genetic trait that may be related to gender. He selects a representative sample of males and females to investigate the question.

We state the null and alternative hypotheses as follows:

H_0: Color blindness is independent of gender (color blindness and gender are not related).

H_A: Color blindness is not independent of gender (color blindness and gender are related).

△ 23.4 DATA INPUT, ANALYSIS, AND INTERPRETATION OF OUTPUT: FIRST EXAMPLE

Following is a 2 × 2 contingency table that shows the data collected by the researcher: 1,000 individuals, 480 males and 520 females, and whether they are color-blind (see Figure 23.1). The numbers in parentheses are expected values and will be described later.

Figure 23.1 Observed and Expected Values for Color Blindness

	Color-Blind	Not Color-Blind	Total
Male	38 (21.12)	442 (458.88)	480
Female	6 (22.88)	514 (497.12)	520
Total	44	956	1,000

By eyeballing the data, are you able to come to any conclusions? You can use SPSS to analyze such data and, after setting an alpha, test the null hypothesis that the two criteria of classification, gender and color blindness, are independent (not related). Is the occurrence of color blindness the same for men and women? If you assume no difference, you can compute (determine) the expected frequencies for men and women, and then test the null hypothesis.

Although SPSS computes the degrees of freedom and expected values for you, it is informative to provide a brief description of these important concepts.

Expected Frequencies: How do we determine expected frequencies for the table shown in Figure 23.1? Under the null hypothesis, we would expect the ratio of color-blind men to all men in the sample to be the same as the ratio of color-blind women to all women in the sample, and also the same as the ratio of all color-blind individuals to the total number of men and women in the sample. Denoting by x the expected number of color-blind men in the sample, we have: $x/480 = 44/1{,}000$ or $x = 21.12$. The shortest method to obtain the expected number of men who are not color-blind is to subtract the number of color-blind men from the total number of men: $480 - 21.12 = 458.88$. You can find the other expected frequencies by subtraction. The expected frequencies are listed in Figure 23.1.

The easy method to obtain expected frequencies is to multiply the values in the margins and divide by the total:

$480 \times 44/1{,}000 = 21.12$

$520 \times 44/1{,}000 = 22.88$

$480 \times 956/1{,}000 = 458.88$

$520 \times 956/1{,}000 = 497.12$

The expected frequencies in Figure 23.1 are shown in parentheses.

Degrees of Freedom: The degrees of freedom (*df*) for a 2×2 table is 1. You can understand this by doing a little experiment. Remove all the frequencies from all the cells leaving only the margin totals. Even with these totals, it is impossible to determine the value for any cell. However, if you place one of the values in a cell, say 38, then you can determine the remaining values for each cell. For example, $480 - 38 = 442$. In general, if the table is $R \times C$, the degrees of freedom can be determined as $(R - 1) \times (C - 1)$. Thus, for a 3×2 table, the degrees of freedom would be $(3 - 1) \times (2 - 1) = 2 \times 1 = 2$.

Let's have SPSS analyze the data in Figure 23.1 to see if color blindness is independent of gender. The challenge is coding the data so SPSS can make sense of it. You must enter variables in SPSS and code these appropriately. Code gender as follows: 1 = *male* and 2 = *female*. Code eye as follows: 1 = *color-blind* and 2 = *not color-blind*. Finally, weight the number (frequency) in each cell. Start SPSS and follow these steps:

Enter the variables:

- In the Variable Screen, enter the variable **gender**.
- Click *Type*, select **numeric,** and set **0** decimal places.
- Click *Values* and enter the following: **1** = **male** and **2** = **female**.
- Click *Measure* and select **nominal**.
- In the Variable View screen, enter the variable **eye**.
- Click *Type*, **select numeric**, and set **0** decimal places.
- Click *Values* and enter the following: **1** = **color-blind** and **2** = **not color-blind**.
- Click *Measure* and select **nominal**.
- In the Variable View screen, enter the variable **number**.
- Click *Type*, select **numeric,** and set **0** decimal places.
- Click ***Measure*** and select **scale**.

Enter the data:

- In the first cell below gender, enter **1**.
- In the first cell below eye, enter **1**.

- In the first cell below number, enter **38**.
- In the second cell below gender, enter **1**.
- In the second cell below eye, enter **2**.
- In the second cell below number, enter **442**.
- In the third cell below gender, enter **2**.
- In the third cell below eye, enter **1**.
- In the third cell below number, enter **6**.
- In the fourth cell below gender, enter **2**.
- In the fourth cell below eye, enter **2**.
- In the fourth cell below number, enter **514**.

Now that you have entered variables, values, and data you can proceed to the analysis.

- In the Main menu, click **Data**, and then click **Weight Cases**.
- A window titled Weight cases will open.
- Click **Weight Cases by**.
- Click **Number** in the left panel, and click the right arrow to place it in the Variable box.
- Click **OK**.
- Click **Analyze**, select **Descriptive Statistics**, and click **Crosstabs**.
- Click **gender** and then click the right arrow to place it in the Rows box.
- Click **eye** and click the right arrow to place it in the Columns box.
- Click **Statistics** and a window will open titled Crosstabs statistics.
- Click **Chi-square** and then click **Continue**.
- Click **Cells**, then click **observed** and **expected**, and then click **Continue**.
- Click **OK**.

Figure 23.2 shows the observed and expected values computed by SPSS, which agree with the values shown in Figure 23.1.

Figure 23.2 Observed and Expected Values for Color-Blind and Not Color-Blind

gender * eye Crosstabulation

			eye		
			color-blind	not color-blind	Total
gender	male	Count	38	442	480
		Expected Count	21.1	458.9	480.0
	female	Count	6	514	520
		Expected Count	22.9	497.1	520.0
Total		Count	44	956	1000
		Expected Count	44.0	956.0	1000.0

Figure 23.3 shows the results of the chi-square test for color-blind and not color-blind.

Figure 23.3 Chi-Square Test for Color-Blind and Not Color-Blind

Chi-Square Tests

	Value	df	Asymp. Sig. (2-sided)	Exact Sig. (2-sided)	Exact Sig. (1-sided)
Pearson Chi-Square	27.139[a]	1	.000		
Continuity Correction[b]	25.555	1	.000		
Likelihood Ratio	29.773	1	.000		
Fisher's Exact Test				.000	.000
Linear-by-Linear Association	27.112	1	.000		
N of Valid Cases	1000				

a. 0 cells (.0%) have expected count less than 5. The minimum expected count is 21.12.

b. Computed only for a 2x2 table

Since .000 is less than .05, the null hypothesis is rejected, meaning that there is a relation between color blindness and gender.

23.5 RESEARCH SCENARIO AND TEST SELECTION: SECOND EXAMPLE △

A researcher is aware that a proposal of national interest, if passed, would involve a possible financial burden on voters. Consequently, he decides to determine if voters with higher incomes would vote differently on the proposal than voters with lower incomes. Since the level of measurement would be nominal and involve only frequencies, the researcher decides that a chi-square test of independence would be appropriate.

23.6 RESEARCH QUESTION AND NULL HYPOTHESIS: SECOND EXAMPLE △

Are voter preference and level of income related?

We state the null and alternative hypotheses:

H_0: There is no difference in the vote insofar as income level is concerned (voter preference and income level are not related).

H_A: There is a difference in the vote insofar as income level is concerned (voter preference and income level are related).

△ 23.7 DATA INPUT, ANALYSIS, AND INTERPRETATION OF OUTPUT: SECOND EXAMPLE

Let's give SPSS another opportunity to analyze the data.

On a proposal of national interest, persons in the two categories (Higher Income and Lower Income) cast votes as indicated in Figure 23.4. At a level of significance of .05, test the hypothesis that there is no difference in the two groups insofar as the proposal is concerned.

Figure 23.4 Observed and Expected Values for Voting Preference

	In Favor	Opposed	Undecided	Total
Higher Income	85	78	39	202
Lower Income	118	61	24	203
Total	203	139	63	405

In the Variable View screen, code the variables as follows.

- Start SPSS.
- In the Variable View screen, enter the following three variables: **Income**, **Vote**, and **Total**.
- Code the variables as follows: Higher Income = **1** and Lower Income = **2**. In Favor = **1**, Opposed = **2**, and Undecided = **3**. Set **0** decimal points for any of the three variables. Set type for each of the three variables as **numeric** with decimals set at **0**. Set measure for income to **nominal**, vote **to nominal**, and total to **scale**.

Enter the following in the Data View screen:

- In the first cell below Income, enter **1**.
- In the first cell below Vote, enter **1**.
- In the first cell below Total, enter **85**.
- In the second cell below Income, enter **1**.
- In the second cell below Vote, enter **2**.
- In the second cell below Total, enter **78**.
- In the third cell below Income, enter **1**.
- In the third cell below Vote, enter **3**.
- In the third cell below Total, enter **39**.
- In the fourth cell below Income, enter **2**.
- In the fourth cell below Vote, enter **1**.
- In the fourth cell below Total, enter **118**.

- In the fifth cell below Income, enter **2**.
- In the fifth cell below Vote, enter **2**.
- In the fifth cell below Total, enter **61**.
- In the sixth cell below Income, enter **2**.
- In the sixth cell below Vote, enter **3**.
- In the sixth cell below Total, enter **24**.

Now that you have entered all variables, values, and data you can proceed with the analysis.

- In the Main menu, click **Data** and then click **Weight Cases**. A window titled Weight cases will open.
- Click **Weight Cases by**.
- Click **Total** in the left panel, and click the right arrow to place it in the Variable box.
- Click **OK**.
- Click **Analyze**, select **Descriptive Statistics,** and click **Crosstabs**.
- Click **Income** and then click the right arrow to place it in the Rows box.
- Click **Vote** and then click the right arrow to place it in the Columns box.
- Click **statistics** and a window will open titled Crosstab statistics.
- Click **Chi-square** and then click **Continue**.
- Click **Cells**, then click **observed** and **expected**, and then click **Continue**.
- Click **OK**.

Figure 23.5 shows the observed and expected values for voting preference.

Figure 23.5 Observed and Expected Frequencies for Voting Preference

Income * Vote Crosstabulation

			Vote			Total
			In Favor	Opposed	Undecided	
Income	Higher Income	Count	85	78	39	202
		Expected Count	101.2	69.3	31.4	202.0
	Lower Income	Count	118	61	24	203
		Expected Count	101.8	69.7	31.6	203.0
Total		Count	203	139	63	405
		Expected Count	203.0	139.0	63.0	405.0

Figure 23.6 Chi-Square Test for Voting Preference

Chi-Square Tests

	Value	df	Asymp. Sig. (2-sided)
Pearson Chi-Square	11.013[a]	2	.004
Likelihood Ratio	11.076	2	.004
Linear-by-Linear Association	10.410	1	.001
N of Valid Cases	405		

a. 0 cells (.0%) have expected count less than 5. The minimum expected count is 31.42.

Figure 23.6 shows the chi-square test results for voting preference. Since 11.01 is greater than .05, the null hypothesis can be rejected at the .05 level. There is a difference in how income groups vote on the proposal.

△ 23.8 SUMMARY

In this chapter, you learned how to use SPSS to analyze data in a chi-square contingency table, how to interpret the output, and how to decide whether to reject or fail to reject the null hypothesis based on the results.

This is the final chapter in this book. At this time, your arsenal and understanding of descriptive and inferential statistics is substantial, as is your understanding of data analysis and which statistical tests are appropriate for your data. Most important, you have a solid command and understanding of SPSS and the various analyses and procedures available to help you progress in your research or simply to give you the tools to understand the research of others.

Appendix A

Data Sets

Table A.1 class_survey1 Variables and Attributes

Variables and Attributes (Properties)									
Name	**Type**	**Width**	**Decimals**	**Label**	**Values**	**Missing**	**Columns**	**Align**	**Measure**
class	numeric	8	0	Morning or Afternoon Class	1 = Morning 2 = Afternoon	None	8	left	nominal
exam1_pts	numeric	8	0	Points on Exam One	None	None	8	left	scale
exam2_pts	numeric	8	0	Points on Exam Two	None	None	8	left	scale
predict_grde	numeric	8	0	Student's Predicted Final Grade	1 = A 2 = B 3 = C 4 = D 5 = F	None	8	left	nominal
gender	numeric	8	0	Gender	1 = Male 2 = Female	None	8	left	nominal
anxiety	numeric	8	0	Self-rated Anxiety Level	1 = Much anxiety 2 = Some anxiety 3 = Little anxiety 4 = No anxiety	None	8	left	ordinal
rate_inst	numeric	8	0	Instructor Rating	1 = Excellent 2 = Very good 3 = Average 4 = Below average 5 = Poor	None	8	left	ordinal

Table A.2 class_survey1 Data

student	class	exam1_pts	exam2_pts	predict_grde	gender	anxiety	rate_inst
1	1	100	83	1	2	4	2
2	1	50	68	3	2	2	1
3	1	78	68	3	2	2	1
4	1	50	78	3	1	2	1
5	1	97	74	2	2	3	2
6	1	41	71	3	2	2	1
7	1	30	72	3	1	1	2
8	1	31	83	2	1	1	1
9	1	71	63	2	2	2	1
10	1	85	89	1	2	3	1
11	1	86	93	2	2	2	1
12	1	67	64	2	2	1	1
13	1	52	100	2	1	2	1
14	1	88	83	1	2	4	1
15	1	25	23	1	1	1	1
16	1	100	100	1	2	2	2
17	1	14	71	3	2	2	2
18	1	60	75	3	2	1	2
19	2	93	84	1	2	2	1
20	2	94	93	1	1	4	1
21	2	90	89	1	1	2	2
22	2	78	80	2	1	2	1
23	2	50	84	3	2	1	1
24	2	74	50	2	2	3	1
25	2	62	93	2	1	4	1

student	class	exam1_pts	exam2_pts	predict_grde	gender	anxiety	rate_inst
26	2	80	81	2	2	2	1
27	2	87	97	1	1	2	1
28	2	25	61	3	2	1	1
29	2	50	82	2	1	2	1
30	2	99	93	2	2	3	1
31	2	50	64	3	2	1	1
32	2	100	100	1	2	2	1
33	2	66	62	3	2	2	1
34	2	50	100	3	2	1	1
35	2	100	94	1	2	4	1
36	2	26	53	4	2	1	3
37	2	41	72	4	2	1	1

APPENDIX B

BASIC INFERENTIAL STATISTICS

△ B.1 INTRODUCTION AND OBJECTIVES

As the term *inferential* implies, one uses inferential statistics to make inferences concerning a defined population of interest. As opposed to descriptive statistics, which describes "what is," inferential statistics predicts what "might be" based on the results of selecting representative or random samples from a defined population of interest.

A researcher may be interested in selecting a sample from the population of males attending a certain university. Defining and describing this population would be relatively straightforward by simply consulting the university records. Another researcher may be interested in investigating the number of cats living in a certain city. Defining and describing this population of cats would be nearly impossible. The bottom line is, it may be easy to "talk" about a population, but unless you are able to define and describe that population adequately, you may have difficulty selecting a representative sample. However, there are methods of selecting samples, although not random, that closely represent a given population.

We introduce some basic concepts related to inferential statistics and describe how SPSS can assist in your analysis of data requiring inferential statistics.

OBJECTIVES

After completing this appendix, you will be able to

- Describe procedures for selecting samples from a defined population
- Describe the process of testing a hypothesis
- Distinguish between Type I and Type II errors
- Describe tests of significance
- Describe Type I and Type II errors and the consequences of each
- Describe degrees of freedom
- Interpret the output of an SPSS *t*-test table
- Distinguish between statistical significance and practical significance

B.2 Populations and Samples

Drawing a random sample from a defined population of interest is the only method to ensure that the sample best represents the actual population. The term *random* does not mean "willy-nilly" as the term might suggest. On the contrary, if a sample is truly random, it means that every member of the population has an equal and independent chance of being selected. However, the procedure for selecting such a sample is often quite difficult or, perhaps, impossible given the size and nature of the population. Fortunately, there are other methods of selecting samples that may obviate this problem.

B.3 Sampling Procedures

Only through random sampling of every element in a population can one assume that the sample is representative of that population. Since such sampling is not always possible, other methods of sampling have been developed, including stratified, cluster, and systematic.

Simple Random Sampling: Selecting a sample such that all individuals in the defined population have an equal and independent chance of selection for the sample. A random procedure such as a random number table or a computer program is used to select the sample. Actually, a computer program generates pseudorandom numbers.

Stratified Sampling: Divides a population into subgroups and obtains samples from each subgroup to assure representation of each subgroup in the sample.

Cluster Sampling: Intact groups, not individuals, are randomly selected.

Systematic Sampling: Every *i*th individual is selected from a list of individuals in a population.

Stratified, cluster, and systematic sampling are not random samples because each member of the defined population dose not have an equal and independent chance of being selected. However, these are used by researchers when no other alternatives are viable. Researches using these methods of sampling often employ inferential statistical procedures to analyze the results. Such results may be open to question because the samples are not actually representative of the total population.

B.4 Hypothesis Testing

In hypothesis testing, two hypotheses are created, the null hypothesis (H_0) and the alternative hypothesis (H_A), only one of which can be true. The null hypothesis states that there is no significant difference in what we observed. Hypothesis testing is the process of determining whether to reject the null hypothesis (the only differences are those due to sampling error) compared with the alternate hypothesis, which states that there is a meaningful difference. Suppose we posit that there is no significant difference between two methods of teaching reading. Any difference we observe is merely the result of random error. If we reject the null hypothesis, it is assumed that the alternate hypothesis is true, meaning that a difference we observe is caused by a systematic difference between groups. For example, a group taught by reading Method A scored significantly higher on the reading test than a group taught by Method B.

B.5 Type I and Type II Errors

When a researcher states a null hypothesis, he is attempting to draw conclusions (make decisions) about a population based on samples drawn from that population. If he rejects a null hypothesis that is true, a Type I error has occurred. If he fails to reject a null hypothesis that is false, a Type II error has occurred. What may be the consequences of Type I and Type II errors?

A school district has decided to try a new method of teaching reading compared with the standard method. Suppose the null hypothesis, which states that there is no difference between these methods of teaching reading as based on students' reading scores, is rejected when it is, indeed, true that there is no difference, indicating a Type I error. The district may then introduce the new method of teaching reading at considerable cost to the district when the efficacy of both methods is the same. Suppose the contrary occurs and we fail to reject the null hypothesis when it is, indeed, false. Then the district will simply continue to use the customary method of teaching at no additional cost. However, the district would be withholding the use of a method of teaching reading that is superior to the customary method.

Most researchers would consider a Type I error the more costly, so they would wish to minimize the chance (probability) of making such an error.

B.6 Tests of Significance

A test of significance is not indicative of "importance" but refers to a statistical level of probability. Significance tests determine the probability of making a Type I error (a preselected probability level we are willing to take if the decision we make is wrong). Typical levels for significance tests are .05 (5 out of 100) and .01 (1 out of 100). A preselected probability level known as level of significance (denoted as alpha, symbol α) serves as a criterion to determine whether to reject or fail to reject the null hypothesis. If, after performing a statistical analysis, a researcher obtains a probability .05 or less, he or she will, by convention, reject the null hypothesis, meaning that the .05 level or less is the maximum Type I error rate the researcher is willing to accept.

Each statistical test in SPSS lists the exact level of significance in an output table. If this level is less than the alpha value you have chosen, then you would reject the null hypothesis. If the alpha value is larger, you would fail to reject the null hypothesis.

B.7 One- and Two-Tailed Tests

If the researcher has evidence that a difference would only occur in one direction, then a one-tailed test of significance would be appropriate. However, if the researcher has no evidence of direction, then a two-tailed test of significance is appropriate. Each statistical test in SPSS lists results for a two-tailed test. If the test is actually one tailed, it is easier to obtain a significant difference. If α is .05 for a one-tailed test, this region is divided into two regions of .025 in a two-tailed test to cover two possible outcomes.

△ B.8 Degrees of Freedom

The term *degrees of freedom* is defined as the number of observations free to vary around a parameter. In short, the number of degrees of freedom is the maximum number of variates that can freely be assigned before the rest of the variates are completely determined.

Let's consider a concrete example to help clarify the concept of degrees of freedom. Consider the following: 2 + 4 + 6 + ? = 20. You can determine the missing number by subtracting the sum of the first three numbers from 20, yielding 20 – 12 = 8. There are no degrees of freedom. Now consider the following: 2 + 4 + ? + ? = 20. There are an unlimited combination of numbers the sum of which is 14, including 7 + 7, 6 + 8, 9 + 5, 3.3 + 10.7, and so on. However, if you choose, for example, the number 11 as one of the missing numbers, then the other number is determined and, in this case, it is 3. Consequently, in this example, there is one degree of freedom.

Each statistical test of significance in SPSS has a particular formula for determining the degrees of freedom. You will see this indicated in output tables as *df*.

△ B.9 SPSS Reports of Hypothesis Testing

The topics presented up to this point are intended to help you interpret the many statistical tables SPSS generates when you request a statistical analysis. Let's use the class_survey1 you saved and compare the means of exam1 points and exam2 points. Suppose the instructor is interested in determining if there is a significant difference in these means using the .05 level of significance. You will use a paired-samples *t* test because you are comparing values from the same group at different times.

- Start SPSS.
- Click **File**, select **Open**, and Click **Data**.
- Locate the **class_survey1** file and open it.
- On the Main menu, click **Analyze**, select **Compare Means**, and then click **Paired-Samples T Test**.
- Click **Points on Exam Two** in the left panel, and click the right arrow to place it in the Paired Variables panel.
- Click **Points on Exam One** in the left panel, and click the right arrow to place it in the Paired Variables panel.
- Click **OK**.

Three tables are generated by SPSS. We show only the third table that lists the results of the paired-samples *t* test as shown in Figure B.1.

Figure B.1 Results of Paired-Samples *t* Test

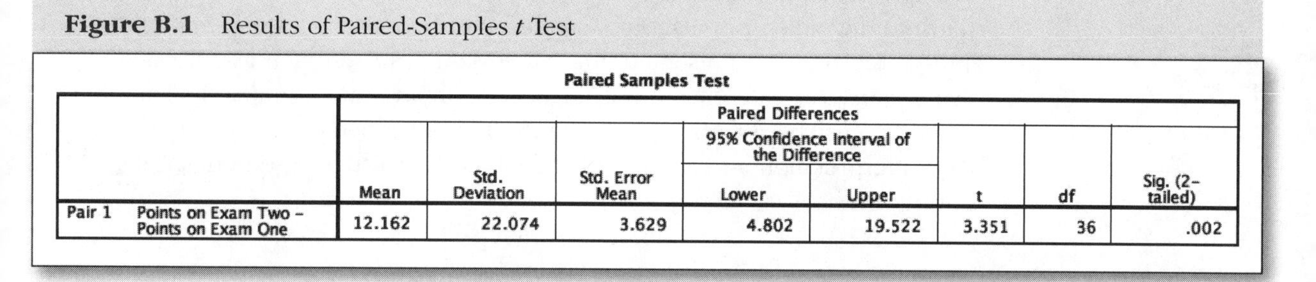

Paired Samples Test

		Paired Differences							
					95% Confidence Interval of the Difference				
		Mean	Std. Deviation	Std. Error Mean	Lower	Upper	t	df	Sig. (2-tailed)
Pair 1	Points on Exam Two - Points on Exam One	12.162	22.074	3.629	4.802	19.522	3.351	36	.002

The value of *t* is 3.351. Degrees of freedom (*df*) is 36, one less than the number of scores in each group. Most important, examine the significance result of .002, which is far less than our α value of .05. Hence you reject the null hypothesis of no difference in the mean scores. There is a difference, and it is due to an actual effect, not chance. Note that this is a two-tailed significance level.

B.10 PRACTICAL SIGNIFICANCE VERSUS STATISTICAL SIGNIFICANCE

The results of a test of significance may be statistically significant, but the actual difference may be minimal and of little importance. For example, a researcher may find a statistically significant difference in two methods of teaching reading as determined by students' scores on a reading test, but the actual difference in these scores may be so minimal that it would not benefit a school district to change methods of teaching reading.

An effect size is a measure permitting a judgment of the relative importance of a difference or relationship by indicating the size of the difference. In statistics, an effect size is a measure of the strength of the relationship between two variables in a statistical population, or a sample-based estimate of that quantity. An effect size calculated from data is a descriptive statistic that conveys the estimated magnitude of a relationship without making any statement about whether the apparent relationship in the data reflects a true relationship in the population. In that way, effect sizes complement inferential statistics such as *p* values.

Four methods of computing effect size are Pearson's *r* correlation, Cohen's *d*, Eta Squared, and Cramer's V. Although SPSS does not provide direct calculations of these effect sizes, you can use output provided by SPSS to compute these values.

△ B.11 Summary

You have been introduced to basic concepts regarding the topic of inferential statistics, including populations, samples, and hypothesis testing. And you interpreted the data presented in a paired-samples *t* test. You also learned that a statistical difference does not necessarily indicate a practical significance. You will be able to apply the concepts you learned in this appendix in your interpretation and analysis of the tables and results based on various statistical tests.

Appendix C

Statistical Tests

C.1 Introduction and Objectives △

SPSS offers two types of statistical hypothesis tests—parametric and nonparametric. The appropriate type of test for your data depends not only on the level of measurement of that data (nominal, ordinal, or scale) but also on the assumptions regarding populations and measures of dispersion (variance). Parametric tests are used to analyze interval (scale) data, whereas nonparametric tests are used to analyze ordinal (ranked) data and nominal (categorized) data.

Parametric tests are those that assume a certain distribution of the data (usually the normal distribution), an interval (scale) level of measurement, and equality (homogeneity) of variances. Probably, the most critical assumption regarding parametric tests is that the population from which a random sample is selected has a normal distribution. For example, if we select a sample from the population of all men living in a certain city and compute the mean height, the assumption is that height is a variable normally distributed in that population of men.

Many parametric tests such as the t test are quite robust, meaning that even though assumptions may not be met, the results of the test are still viable and open to statistical interpretation. In other words, moderate violations of parametric assumptions have little or no effect on substantive conclusions in most instances.

We describe some SPSS procedures you can use to make determinations concerning the normality of populations. In Chapter 11, you were introduced to the concepts of skewness, kurtosis, and standard error. You can use measures of these as a rule of thumb to make a determination regarding normality of populations.

As you work through the subsequent chapters in this book, we will describe various parametric tests, including assumptions related to these tests. We follow some of these parametric tests with a description of a non-parametric test that may be used in their place in case assumptions for the parametric tests cannot be met.

OBJECTIVES

After completing this appendix, you will be able to

Distinguish between parametric and nonparametric tests

Describe and explain the assumptions required for a parametric test

Determine whether a parametric or nonparametric test is appropriate for variables measured at the nominal, ordinal, and interval levels

Describe and explain SPSS procedures available for determining if a variable is normally distributed and if variances are equal, or at least similar

Describe and explain the purpose of data transformations

Use SPSS to transform a variable

△ C.2 Parametric Statistical Tests and Related Assumptions

On reviewing the material presented in Chapter 11, you will recall that skewness is a measure that indicates symmetry of a distribution. When a distribution is not normal, it is said to be skewed. A normal distribution is bell-shaped with a mean of 0 and standard deviation of 1. In a normal distribution, the mean, median, and mode are identical. Skewed distributions have a characteristic shape. When a distribution is positively skewed, many of the scores are bunched at the bottom of the range of scores, while a smaller number of scores extend toward the top of the range. In a positively skewed distribution, the mode is less than the median, and the median is less than the mean: mode < median < mean. In a negatively skewed distribution, the opposite occurs: mean < median < mode.

Let's use an SPSS procedure to examine the mean, median, and mode of points on exam one in the class_survey2 database you saved in Chapter 6.

- Open **class_survey2**.
- Click **Analyze**, select **Descriptive Statistics**, and then click **Frequencies**.

- **Click Average Points On Exams** in the left panel and click the right arrow to place it in the Variable(s) box.
- Click **Statistics**, and then click **Mean**, **Median**, and **Mode**. Unclick any other items that may be selected.
- Click **Continue** and then click **OK**.

The results are shown in Figure C.1. The mean = 72.0, the median = 73.0, and the mode = 57.0. Actually, there are three modes in this distribution, which you can see by inspection of the following graph in Figure C.2. When there are multiple modes, SPSS shows the smallest. Notice that the mean and median are quite close in value indicating that the distribution appears to represent a normal distribution.

Figure C.1 Mean, Median, and Mode on Average Points on Exams

Statistics

Average Points On Exams

N	Valid	37
	Missing	0
Mean		72.03
Median		73.00
Mode		57[a]

a. Multiple modes exist. The smallest value is shown

Let's create a graph of this distribution and superimpose the normal distribution over it as you did in Chapter 8 so that we may inspect (eyeball) the distribution to see if it seems representative of a normal curve.

- Open **class_survey2**.
- Click **Graphs**, select **Legacy Dialogs**, and then click **Histograms**. A window will open titled Histogram.
- Click **Average Points On Exams** in the left panel, and click the right arrow to place it in the Variable box.
- Click **Display normal curve**.
- Click **OK**.

Figure C.2 Average Points on Exams

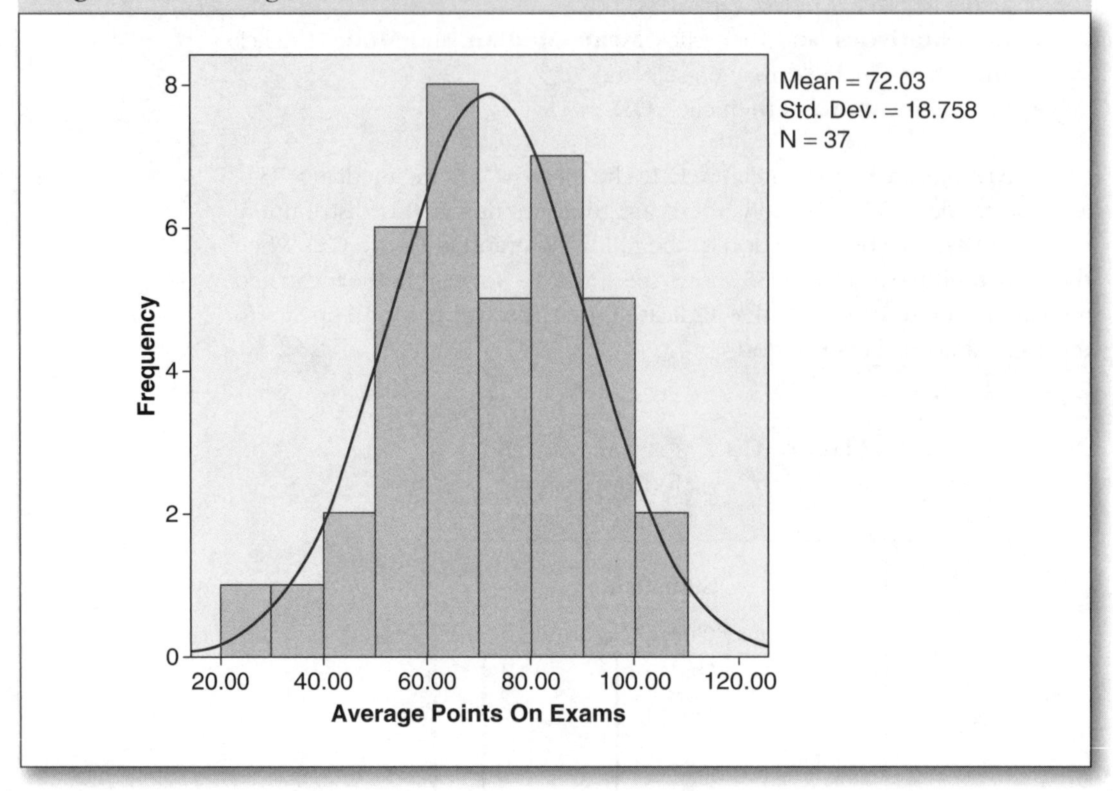

When inspecting this graph, it appears that the histogram fits the normal curve quite closely and could reasonably represent a normal distribution. If values were clustered far to the right or left, or both, then you would not be looking at a normal distribution. Does SPSS offer any additional procedures to help us decide if the distribution is normal?

A common rule-of-thumb test for normality is to run descriptive statistics to obtain skewness and kurtosis for a variable, and then divide these by the standard error. Skew and kurtosis should be in the +2 to −2 range when the data are normally distributed. Don't worry, you do not have to compute these statistics, because SPSS will do it for you! See Chapter 11 for a discussion of standard error, kurtosis, and skewness.

Let's let SPSS compute the standard error, kurtosis, and skewness for the variable **Average Points On Exams** used for the graph in Figure C.2 to see if we can make a determination concerning whether this variable seems to fit the normal curve.

- Open **class_survey2**.
- Click **Analyze**, click **Descriptive Statistics**, and then click **Frequencies**.
- Click **Average Points On Exams** and then click the right arrow to place it in the Variables box.
- Click **Statistics**.
- Click **Skewness** and **Kurtosis** and unclick any other items that may be selected.
- Click **Continue** and then click **OK**.

The results are shown in Figure C.3.

Figure C.3 Standard Error, Kurtosis, and Skewness

Statistics

Average Points On Exams

N	Valid	37
	Missing	0
Skewness		-.451
Std. Error of Skewness		.388
Kurtosis		-.291
Std. Error of Kurtosis		.759

Divide −0.451 by 0.388 and the result is −1.16

Divide −0.291 by 0.759 and the result is −0.383

Using the rule of thumb, both results indicate that the data for this variable follow a normal distribution, which is what our visual inspection of the graph indicated.

Parametric tests should not be used unless the following have been determined: (1) samples are random, (2) values are independent, (3) data are normally distributed, and (4) the samples have equal or at least similar variances. However, researchers often take some latitude in applying these constraints. For example, if using a Likert-type scale in which each response is assigned a point value, say from 1 to 5, researchers often assume an interval level of measurement among the items, when, in actuality, the Likert-type scale should be considered as ordinal data, because one cannot show that the distance between a choice of 1 and 2 is equal to the distance between 2 and 3, and so on. When such decisions are made, one must critically consider the outcomes of research using such scales.

Also, in practice, levels of measurement are often downgraded from scale to ordinal or nominal, mainly for convenience and use of a more familiar test such as chi-square. This is not really good practice, because one is casting out information: There is more information contained in interval data than ordinal data.

The *power* of a statistical test is defined as its ability to detect a significant difference in a sample of data. The power of a parametric test is greater than that of a nonparametric test. Consequently, one should use a parametric test whenever assumptions such as normality of distributions and equivalences of variance can be met.

△ C.3 TRANSFORMATIONS

Data transformation of values for a variable is obtained by applying mathematical modifications. If variables are not normally distributed, you may apply nonparametric tests or you may change the scale of measurement using a transformation. There are a host of transformations available, and many of these simply reduce the relative distances between data points to improve normality. Three important transformations are the Square Root Transformation, the Logarithmic Transformation, and the Inverse (Reciprocal) Transformation. For example, if you apply a Square Root Transformation to your data, the square root of each value is computed.

If you ever require a transformation of your data, SPSS can handle it easily. You will use the database with which you are familiar to perform a Square Root Transformation of data values for the variable **exam1_pts**. Actually, this variable does not require a transformation, but we wish to show you how to do it in case a situation arises with your actual data that may require a transformation.

- Open **class_survey2**.
- Click **Transform** and then click **Compute variable**. A window titled Compute Variable will open.
- In the Target Variable box type **sqrtexam1**.
- In the panel titled Function Group on the right, click **Arithmetic**.
- In the lower panel titled Functions and Special Variables, scroll down and click **Sqrt**.
- Click the upward pointing arrow to place Sqrt in the Numeric Expression box.

- In the left panel, click **Points on Exam One** and then click the right arrow to place it in the parentheses following SQRT in the Numeric Expression box.

Your screen should look like that in Figure C.4.

Figure C.4 Compute Variable Screen With sqrtexam1 as New Variable

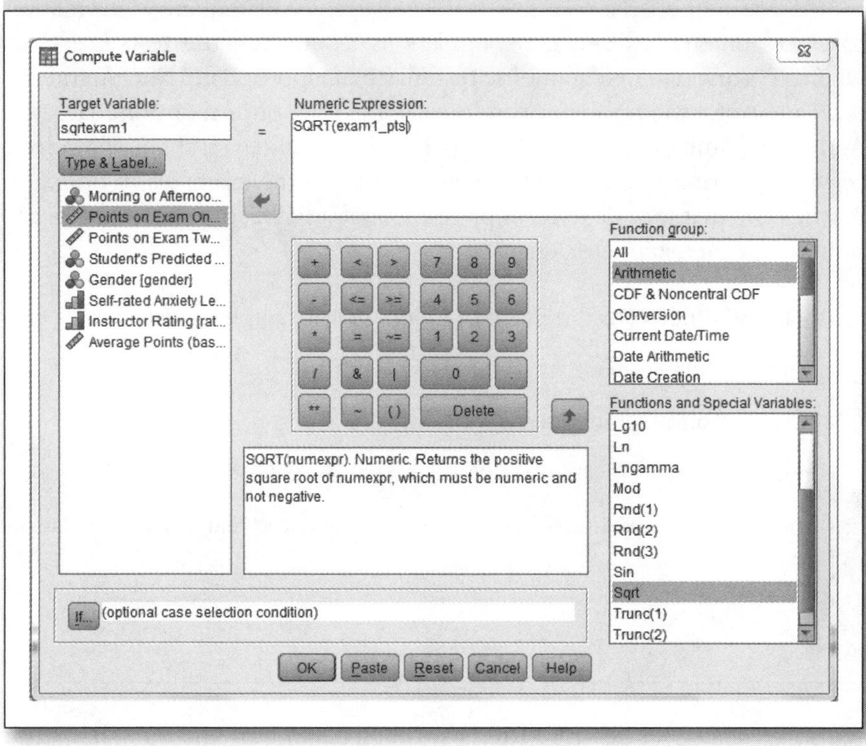

- Click **OK**.

A new variable titled sqrtexam1 will appear in the Data View screen with the square root values listed. At this point, you could use SPSS to perform any analysis on these data that may be required.

C.4 NONPARAMETRIC STATISTICAL TESTS AND RELATED ASSUMPTIONS △

In Section C.3, you learned how to use SPSS to perform data transformations. At times, such transformations may be useful. However, there is another option, namely, the use of nonparametric statistics to analyze your data. Assumptions are usually quite minimal when using nonparametric tests.

Nonparametric tests are designed for databases that may include counts, classifications (nominal), and ratings. Such tests may be easier for a layperson to understand and interpret.

Statistical analyses that do not depend on the knowledge of the distribution and parameters of the population are called nonparametric or distribution free methods. Nonparametric tests use rank and frequency information to assess differences between populations. Nonparametric tests are used when a corresponding parametric test may be inappropriate. Nonparametric tests are useful when variables are measured at the ordinal or nominal level. And nonparametric tests may be applied to interval data if the assumptions regarding normality are not met. Following are some examples of nonparametric tests that may be used in place of parametric tests if assumptions are not met for the parametric tests.

Mann-Whitney *U* test in place of independent-samples *t* test

Wilcoxon test in place of paired-samples *t* test

Kruskal-Wallis test in place of one-way ANOVA

Friedman test in place of repeated-measures ANOVA

Spearman's correlation coefficient in place of Pearson's correlation coefficient

△ C.5 SUMMARY

In this appendix, we described and explained parametric and nonparametric statistical tests and provided the rationale and criteria for deciding which may be appropriate for analysis of your data. While there are assumptions including normal distributions of data and, possibly, equivalences of variances, parametric tests offer more power compared with their nonparametric equivalents, meaning that a parametric test may find a significant difference in a test of hypotheses, whereas a nonparametric test will not when applied to the same data.

Alternatives exist when applying statistical tests to your data. If appropriate, you may wish to transform that data via a mathematical transformation to produce a variable more closely resembling a normal distribution. Or you may prefer to apply appropriate nonparametric statistical tests.

Appendix D

Analysis of Covariance: Test for Homogeneity of Regression Slopes

We mentioned in Chapter 18, which described the analysis of covariance (ANCOVA), that the assumption of "homogeneity of regression slopes" is very important when it comes to interpreting the results of an ANCOVA. And we also mentioned that we had conducted a test, using SPSS, and determined that the assumption of homogeneity of regression slopes was not violated in the example we presented. Since ANCOVA is used by many researchers, we feel it is imperative that the reader be provided with the SPSS analysis to test this assumption.

Saying that one assumes homogeneity of regression slopes simply means that one assumes that the relationship between the dependent variable, the posttest scores in our example, and the covariate, the pretest scores in our example, is the same in each of the treatment groups, the four methods of teaching reading. For example, if there is a positive relationship between the pretest scores and the posttest scores in Reading Method one, we assume that there is a positive relationship between the pretest scores and the posttest scores for Reading Methods Two, Three, and Four. Another way to state this is as follows: the regression lines, when pretest scores are plotted against posttest scores for each of the four methods of teaching reading, are essentially parallel, meaning each has the same slope. SPSS can produce graphs of these regression lines if you wish, but this is beyond the scope of our book.

In order to have SPSS test for homogeneity of regression slopes, we will repeat the ANCOVA test used in Chapter 18, but in this case we will request a custom analysis that will enable us to request a test of the interaction between reading method and pretest scores. The results of the interaction between pretest scores and posttest scores will answer the question regarding homogeneity of regression slopes.

Follow these steps:

- Start SPSS.
- Click **Analyze**, select **General Linear Model**, and then click **Univariate** and a window titled Univariate will open.
- Click **Posttest Points** and click the right arrow to move it to the Dependent Variable box.
- Click **Reading Method** and click the right arrow to move it to the Fixed Factor(s) box.
- Click **Pretest Points** and click the right arrow to move it to the Covariate(s) box.
- Click **Model**.
- Click **Custom**.
- In the Build Terms Type box, scroll and select **Main Effects**.
- In the Factors and Covariates panel, select both **ReadMethod** and **Pretest** and move these to the **Model** panel.
- In the Build Terms Type box, scroll and select **Interaction**.
- In the Factors and Covariates panel, hold the **Ctrl** key down and simultaneously select both **ReadMethod** and **Pretest** and move these to the Model panel.

At this point, your screen should look like that shown in Figure D.1.

Figure D.1 Test for Homogeneity of Regression Slopes

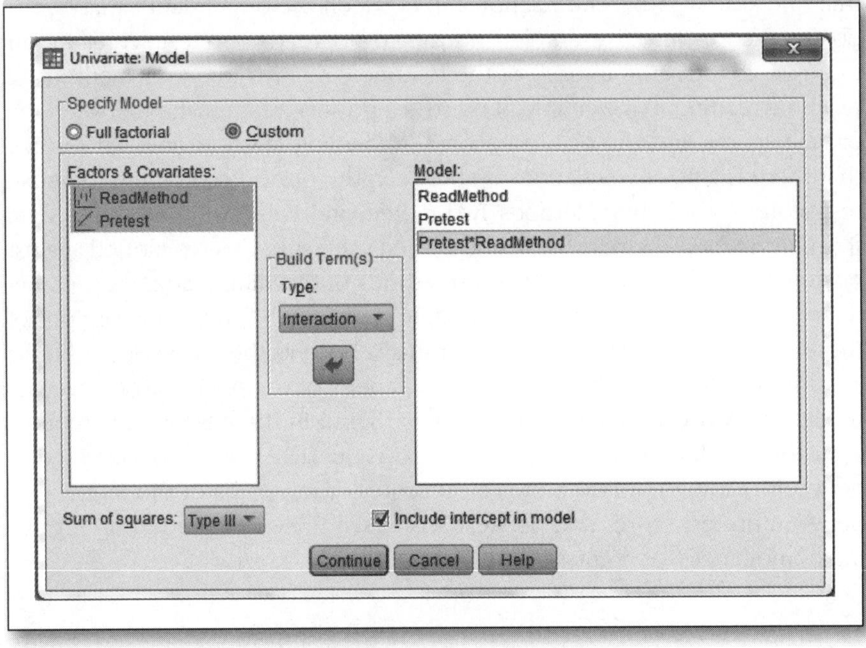

- Click **Continue** and then click **OK**.

Figure D.2 shows the results of the test for homgeneity of regression slopes. On inspecting the ReadMethod*Pretest row, we find that the level of significance is .340. Since .340 is greater than .05, we fail to reject the null hypothesis that there is no interaction, indicating that the regression slopes are homogeneous.

Figure D.2 Results of Test for Homogeneity of Regression Slopes

Tests of Between-Subjects Effects

Dependent Variable:Posttest Points

Source	Type III Sum of Squares	df	Mean Square	F	Sig.
Corrected Model	7194.115[a]	7	1027.731	18.468	.000
Intercept	3253.904	1	3253.904	58.472	.000
ReadMethod	216.960	3	72.320	1.300	.309
Pretest	3193.681	1	3193.681	57.390	.000
ReadMethod * Pretest	200.931	3	66.977	1.204	.340
Error	890.385	16	55.649		
Total	659506.000	24			
Corrected Total	8084.500	23			

a. R Squared = .890 (Adjusted R Squared = .842)

Index

Note: In page references, f indicates figures.